THE CORPOREAL LIFE OF SEAFARING

DISCOURSE 011

THE CORPOREAL LIFE
OF SEAFARING

LALEH KHALILI

For my Alfred

Human beings are magical. Bios and Logos. Words made flesh, muscle and bone animated by hope and desire, belief materialized in deeds, deeds which crystallize our actualities.

Sylvia Wynter, "The Pope Must Have Been Drunk, the King of Castile a Madman"

But the body is also directly involved in a political field; power relations have an immediate hold upon it; they invest it, mark it, train it, torture it, force it to carry out tasks, to perform ceremonies, to emit signs. This political investment of the body is bound up, in accordance with complex reciprocal relations, with its economic use [...].

Michel Foucault, *Discipline and Punish*

When I first travelled on a freighter, the shipping agent who arranged my travel booked me into a seafarers' hotel in the port of Marsaxlokk, Malta, about ten miles southeast of Valletta, the Maltese capital. I sometimes daydream while looking at maps; with its mixture of Arabic, Maltese, and Italian names, the map of Malta reminds me of place names in the fantastical fables of Jorge Luis Borges. Marsaxlokk's name is a combination of the Arabic *marsa* (or anchorage) and *xlokk*, the Maltese name for sirocco or *khamsin* winds blowing north from the Sahara. The names on the map of Malta hint at this island being a palimpsest of maritime flows, trade, and imperial conquests – the marriage of Mediterranean commerce and war.

Malta has long been a strategic port for both commercial and naval vessels, foremost among them those of Britain, for fuelling, victualling, and repair. Malta's island location, equidistant from European and North African coasts, and its extant infrastructures, including its character as something of an offshore haven *within* the European Union, make it a perfect transhipment port for both licit and illicit cargo. Transhipment ports receive massive container ships capable of carrying 24,000 containers, or more than 200,000 tonnes of goods, and distribute their cargo to nearby coasts via smaller feeder ships. The behemoths then sail to the industrial ports of northwestern Europe or towards East Asian ports that dispatch products manufactured in the factory of the world, China.

I first arrived in Marsaxlokk at the end of January 2015, planning to steam on one of these giants, CMA CGM *Corte Real*. CMA CGM is the world's third largest shipping company, after the Danish Maersk and the Italian Mediterranean Shipping Company, which compete for first place. It is owned by the Lebanese/Syrian Saadé family, who moved to Marseille in 1975 and shortly thereafter set up Compagnie Maritime d'Affrètement, a roll-on/roll-off vehicle carrier company. They soon expanded into container and bulk transport, and eventually acquired the French colonial shipping colossus Compagnie Générale Maritime, privatised in the flurry of 1990s neoliberal reforms in France. Designed by the Iraqi-British architect Zaha Hadid, CMA CGM's stylish headquarters stands proudly above the ferry port in Marseille, but their shipping hub is 700 miles offshore, in Marsaxlokk.

The seafarers' hotel which I was booked into by CMA CGM's agent in Malta loomed over the corniche, its windows overlooking the port. My room was large and sparse, with tile floors that remained cool even in the summer heat, wonky curtains barely covering the window, and a squeaky single bed with a thin mattress and crisp, if threadbare, white sheets. The town itself drowsed in the sun, its peeling walls and boarded-up shopfronts redeemed by the Mediterranean sandstone and sunlight. Some businesses had English names, and most shopkeepers spoke an international argot that incorporated English, Italian, and Maltese. I understood a little Maltese because of all its Arabic loanwords.

In the mornings, when I came down to the spartan dining room for breakfast, the seafarers on their way home after four- or nine-month stints at sea were often hungover. Those arriving to the hotel from the nearby airport to board ships for similar stints were restless and anxious. The seafarers gazed at me curiously whenever they encountered me; I imagine they must have wondered what I was doing there amongst them, this lone woman older than most of them by a decade or two, and definitely a landlubber. They mostly left me alone.

CMA CGM *Marco Polo* in Marsaxlokk, Malta

The men I met at that hotel, either Filipino or from former Yugoslavia, were quiet and polite, mostly in their twenties and thirties. They socialised with their own countrymen, speaking in Croatian or Tagalog. As I later learned aboard ship, nationality structures seafarer conviviality in ways that social class, rank, or ratings (wages) do not. National attachments also decide whether one goes to sea on coastal vessels or ocean-going ships. The Philippines depends on remittances from ocean-going maritime workers. China, Russia, India, and Indonesia are not far behind in proffering seafarers to the global shipping industry.[1] All littoral countries produce their own maritime workers, though many – including the *dhow* sailors of the Western Indian Ocean – ply their trades in coastal and shorter routes.

In the transition from sail to steam, human power (and numbers) aboard ship shifted from deckhands and sailors to engine room workers. Leon Trotsky called shipboard engineers "industrial workers in sailor's uniform." In that period, engine room workers were often resented, not only by ship owners and officers who considered them unruly, but even by sailors, precisely for outnumbering them aboard ships.[2] In *Ultramarine*, an autobiographical novel based on Malcolm Lowry's experience of going to sea in the 1920s, seafarers constantly remind the greenhorn protagonist about the national apportionment of work on the ship: "Nearly all of us are Norwegian our side of the fo'c'sle, but the two cooks are Norse too – the sailors are nearly all English on yours." When he is told that deckhands and firemen are in solidarity with one another, he cynically thinks, "as though Britons and Norwegians, a Spaniard, an American, and a Greek would spend their watch below in brotherly communion!"[3]

Some of this interconnection between seafarers' nationality and their duties is legislated. For example, the United States' Jones Act requires that any ships carrying cargo between two US ports must be owned by a US shipping firm and carry a majority-American crew. Some of it arises from colonial regimes of labour, as, for example, in the historical predominance of *lascars*, South Asian sailors, on British and Dutch ships – or, more accurately, in their stokeholds. And some of it has to do with the global colour line running through wage and recruitment structures of international enterprises. Seafarers' passports determine a great deal about the kinds of work they do and the kinds of wages they earn.

Three days after arriving in Marsaxlokk I was finally given the signal to board the ship. That was the beginning of the first of two journeys, twenty months apart, from Malta to Jabal Ali, Dubai, on two different CMA CGM ships. The journeys were different in many ways, even if their starting and end points were the same. One took place at a peak in global trade; the ship steamed through the Mediterranean, the Red Sea, the Arabian Sea, the Gulf of Oman, and the Persian/Arabian Gulf at breakneck speed, with very few intermediate stops, only to be forced to anchor off the coast

Containers loaded outside my porthole on *Corte Real*

of Jabal Ali, waiting for a berth to come free. It sped in order to deliver its enormous cargo of chemicals and machinery to Dubai, and thereafter to Southeast and East Asia. Shipping fuel prices were high, but a delay in delivery would have cost even more. The ship was so laden with containers that my porthole had no view to the sea.

The second journey, aboard the CMA CGM *Callisto*, took longer. The ship sailed through four other ports – Damietta, Beirut, Mersin, and Jeddah – before arriving in Dubai. These intermediate destinations were shoehorned into the schedule to pick up new cargo that might justify the ship's fuel expenditures. The ship also went slower to reduce diesel usage. For most of the journey, the containers on deck were sparse, and my seaview was no longer blocked.

Both *Corte Real* and *Callisto* were nearly 400 metres long. Since my second journey in 2016, freighters have become longer still, but also wider, taller, and deeper today than they

Two-way traffic in the Suez Canal in August 2016

once were. *Great Republic*, a clipper launched in 1853 in Boston and used later by the French to transport munitions and troops to Crimea, was never surpassed in size among wooden-hulled sailing ships. After the Crimean War ended, the *New York Times* reported that it had arrived in London "with the largest cargo of guano that ever crossed the ocean in one vessel – over 3,500 tons."[4] *Great Republic* was almost 110 metres long, had a draught of nearly eight metres, and could reach nearly twenty knots. When steamships overtook sailing vessels, the composition of crew members on board also changed. Deckhands and mates were soon outnumbered by firemen, trimmers (shipboard coal carriers), and, as mentioned above, engineers working in the stokehold. A typical ship could have nearly thirty seafarers working below deck, in three shifts of ten men each, while deckhands on the same ship could number around twenty. The *Titanic*, which was the largest steamship of its generation, reached 270 metres long, and had 280 men working below decks in shifts, stoking the engine, trimming the coal, and lubricating the moving parts.

The Port of Beirut before the 2020 Explosion

Inventions in ship design and advances in materials mean that ships have become larger since the *Titanic*, but not as long as they can be. The reasons for this are both physical and economic. A ship can be so long that when caught between two oceanic peaks, its fore and aft lift and its middle sags. This can lead to fissures in the hull, and the strain of the torsion can shear the ship in two. Between peaks in global trade, freighters steam below capacity and are unable to recuperate the cost of crewing, fuel, and transit fees. Large ships ordered when the market is booming may have to be scrapped in economic downturns, a scant handful of years later. Ship sizes grew particularly dramatically when two wars shut the Suez Canal in 1958 (for eight months) and 1967 (for eight years), severing the flow of crude from the oil fields of the Middle East to European refineries and factories via the canal. Shipowners ordered ever more enormous crude tankers to meet the economies of scale, rounding the Cape of Good Hope with larger cargoes of oil. Merchant Tanker *Seawise Giant*, laid down in 1979 and scrapped thirty years later, was nearly 460 metres long, sixty metres longer than today's largest container ships.

Despite these ballooning ship sizes, fewer seafarers operate today's behemoths than older ships a quarter their size. CMA CGM *Corte Real* had thirty-one crew members, including three trainee cadets; *Callisto* had just twenty-six. *Corte Real* also carried the captain's wife and two other women passengers (one of the cadets was also a young Croatian woman). I was the only woman and the only passenger on the second ship.

Today's technologically sophisticated freighters do not, however, steam much faster than their predecessors. As the American photographer and theorist Allan Sekula writes,

> Acceleration is not absolute: the hydrodynamics of large-capacity hulls and the power output of diesel engines set a limit to the speed of cargo ships not far beyond that of the first quarter of [the twentieth] century. It still takes about eight days to cross the Atlantic and about twelve to cross the Pacific. A society of accelerated flows is also in certain key aspects a society of deliberately slow movement.[5]

While time spent at sea may be unexpectedly consistent across eras and ships' fuels, turnaround times at ports have shrunken. In times past, bulk deliveries required scores of stevedores to unload the vessel, and ships could remain in dock for a week or more. Walter Benjamin, who found freighter travel seductive, describes his experience of arrival at port: "Now I'm lying on deck, the evening in Genoa before me, and the sounds of unloading freighters all around me as the modernized 'music of the world.'"[6] He had time at port to explore the city's neighbourhoods, and visited taverns and hidden corners of Spanish and Italian ports alongside the ship's captain and mates, who knew the cities from previous visits. The advent of freight containers changed this.

Today, a twenty-four-hour turnaround time would be considered too long; most port visits by container ships and hydrocarbon carriers take, at most, eighteen. And as ports move further out of city centres, and as the security and identification requirements to depart port grounds into the city become more stringent – thanks in part to US impositions since 2001 – raucous port visits have largely become a thing of the past. With such short turnaround times, seafarers touch land only to use facilities on the grounds of the port – these often belong to charities – and are otherwise effectively captive at sea for months on end. Such a state of confinement was exacerbated by the COVID-19 pandemic. Global limitations on disembarking at ports and closure of airports during the health emergency resulted in 400,000 seafarers in 2020, and 200,000 in 2021, being stranded at sea, unable to go home even months after their contracts had ended.

Are ships "total institutions"?[7] Vilhelm Aubert, a Norwegian sociologist who researched seafarers on mid-century oil tankers, argued that their "remoteness at sea" made ships a "hidden society" comparable to "the privacy of love, the secrecy of the underground, the isolation of the ill [and] the retirement from social life into sleep."[8] Less whimsically, the scholar Vivek Bald, who has written on South Asian seafarers jumping ship and migrating to Harlem, described early twentieth-century seamen as "semicaptive and hyper-exploited but globally mobile."[9] To live on the sea requires that the seafarer's body and soul adapt to weather, to oceanic movements, and to work performed with unorthodox rhythms. Monotony and frenetic activity, banality of routine and skilled repetition, scrupulousness and care, and an uncanny mixture of loneliness and claustrophobia – the affective results of confinement – shape shipboard life.

The quotidian life of labour, tedium, longing, loneliness, and camaraderie aboard ships tracks and illustrates the macro-historical, gargantuan movements of capital, cargo, and people across vast oceans. In earlier eras, seafarers on all shores were pressed into service, and many died at sea to enrich shipowners, kings, and merchants. Early twentieth-century accounts of pearling boats in the Persian Gulf include stories of seafarers forced to dive by the threat of the whip or debt bondage. Many seafarers employed on industrial fishing vessels today are effectively enslaved aboard, and those who resist sometimes disappear at sea.

Coastal and transoceanic trade saw – and still sees – seafarers engaged in backbreaking and lonely work aboard ships because of necessity, obligation, or the desire to engage in petty trade to pay debts or supplement meagre earning.[10] George Hourani's description of early modern seafaring in the Indian Ocean, for example, does not recommend a life at sea:

> Sea voyages in those days were full of hardship. To begin with, the ships were often overcrowded. [...] Then there were lengthy stops in hot, humid ports, where the ship was at the mercy of the local ruler; expensive port dues and presents had to be paid, and there would often be forced stays ending only when the possibilities of trade were exhausted. On the ocean, storms, reefs, and shallows were ever-present perils; captain and crew felt only slightly less helpless than the merchants; in the midst of huge waves man was indeed 'a worm on a splinter.' Add to this the terrible danger of pirates in their oared vessels, much faster in calms and light winds than any ship relying on sail alone; these could be repelled only by the action of fire-throwing marines carried aboard, except in the rare waters where a ruler kept a navy to safeguard shipping.[11]

The great historian of seafaring Marcus Rediker writes of sailing ships traversing the Atlantic in the eighteenth century, and shows these seafarers' work to have been unfree, shaped by violence, and punctuated by turns to mutiny.[12] Black seafarers and *lascars* had their own lacerating experiences of work aboard, with drudgery exacerbated by contract provisions that wrote racialised violence – hunger, pittance-level wages and inadequate accommodations – into extant draconian hierarchies.[13]

Much has changed the character of labour aboard ships: transformations in maritime technologies and the geographic processes of capitalisation; the struggle for collective bargaining and worker representation; and new transnational, state and nongovernmental institutions formed to discipline

Unloading the ship in a port in the UAE

labour, aid capital accumulation, or restrict the worst abuses aboard ships. But much remains constant, especially strict hierarchies aboard ocean-going ships, maintained through variant contract conditions between officers and crew, and through ceremonial and routine shipboard practices.

The Filipino seafarers on my two journeys were on long contracts: nine months onboard ship and one month off. The Croatian officers, by contrast, had four-month-long contracts. Captains helmed the ship for just two months. On the first ship, the captain was permitted to bring his wife along, a privilege afforded certain officers employed by shipping companies rather than working on contracts. It also became clear that many of the Filipino, Chinese, and Indian men came not from port cities but hinterland agricultural towns and villages. This was also the case with Yemeni and Bengali *lascars* aboard older European ships.[14] In an earlier age, accumulation of debt on farms, a bad year for harvests, or simply escaping the strictures of a small town,

The red ensign flying over our ship in Damietta

drove many Arab, African, and Asian seafarers to ports where either local *serangs* (recruiters) or shipping agency recruitment offices enlisted them for service on ships.[15]

Both *Corte Real* and *Callisto* flew the British Red Ensign. The flag under which a ship is registered determines which country's labour, environmental and tax laws the ship will follow, and which country's inspectors will discover and remedy, or punish, a registered ship's misdeeds. The flag country of the ship also sets the minimum level of manning ships, i.e. the number of crew that can safely operate a vessel. While the International Labour Organisation mandates rest-hours stricter than standards set by interstate maritime treaties, specific flag regulations and collective bargaining agreements (if any, and often there are none) often fall short of those standards.

The twentieth-century invention of flags of convenience – or open registries – unmoored a ship's ownership structures from the flag it flew. It was also intended to loosen the hold of tightening labour laws, union scrutiny, environmental inspections, and state tax claims on shipping companies.[16] Their invention was also a function of uneven neo-colonial relations between great commercial powers and smaller states seeking revenue sources. It is no surprise that the ships that flew the flags of the earliest open registries, in Panama and Honduras, were the United Fruit's banana boats and Standard Oil's tankers. The firms that register ships in flag of convenience countries have often been headquartered elsewhere, and in the case of some of the largest flag countries – Panama, Liberia, and the Marshall Islands – have expatriated a substantial portion of the revenues to the United States. The most egregious example is Liberia, where Edward Stettinius Jr., formerly Franklin Roosevelt's Secretary of State and ambassador to the United Nations, set up a registry whose headquarters remain located in Vienna, Virginia. When the Liberia Company was established, sixty percent of the profits recurred to the corporation; ten percent went to a "development fund" Stettinius had set up to encourage investment in extractive industries; only twenty-five percent was given to the Liberian government.[17]

Flagging ships to Panama, Liberia or the Marshall Islands is not only a means of hiding income from taxation, or bypassing inspection of polluting machinery. Significantly, it is a way to pay crew members less. The paucity of regulatory strictures means that shipowners set wages in ways that benefit them. US shipping companies pioneered moving their fleets to flags of convenience, but when the US Supreme Court ruled in 1963 that maritime unions had no say in the decision on how to flag oceangoing ships, the shifting of fleets to open registries accelerated. In 1982, for example, Exxon reported that a tanker carrying twenty-eight seafarers had to pay $2.5 million in wages if flagged to the US, but only $560,000 if flagged to the Philippines.[18] Today more than half the world's goods by tonnage is transported under flags of convenience.[19]

Being under the Red Ensign meant that I was on a relatively well-inspected ship where working conditions were better than most. But it is important to note that even European states with relatively better regulations now have, alongside their national flags, international registries that have laxer rules aboard ships and which offer lower wage scales. The Global Union Federations' struggle against flags of convenience has not been without controversies over the question of transnational asymmetries of power and access to shipboard employment. The International Transport Workers' Federation (ITF) archives at Warwick University, for example, contain correspondence from Indian seafarers' unions complaining that ITF campaigns against flags of convenience often translated into constraints on the employment of Indian seafarers on international ships.

The move to international or open registries has removed limits on the employment of foreign nationals aboard ships, but without legal requirements that those seafarers be paid at the same wage scale as national seafarers. In fact, under flags of convenience, a ship owned by France-based CMA CGM can fly a Panamanian or Liberian flag and pay its seafarers a fraction of what a French seafarer would have demanded. The international registries of European states also aspire to this degree of flexibility and labour arbitrage, but are sometimes limited by residual labour regulations. In fact, the French and German international registries are designated as flags of convenience by the ITF. Ships' manning structures, therefore, starkly distil and display global inequalities. The ships I journeyed on were officered by Croatian officers and crewed mostly by Filipino seafarers. The Croatian officers are less expensive than German or British officers would have been, and their contract terms are less congenial. But they are better paid and treated than the Filipino seafarers. The Philippines now provides the largest percentage of seafarers in the world – a quarter of all seafarers worldwide – and the Philippines depends on the remittance they send home: five percent of all Filipino workers are seafarers, and they provide just under four percent of the country's GDP.[20]

The cosmopolitanism of shipboard life has always pressed against hierarchies that often follow political and geopolitical striations. Historian Valerie Burton records a British captain commenting on "a North Atlantic steamer in 1897 with British officers but 'a Chinese steward, French cook, German carpenter, Greek lamptrimmer, Italian donkeyman; among the A.B.'s two Japanese, two Swedes and a German and among the firemen one Japanese, one West Indian, one French negro and one Portuguese negro.'" Burton reports that this British captain sneered: "Is it likely that any self-respecting Britain (sic) will care to eat, drink, and live for months at a time in a small forecastle with such a motley horde?"[21] By contrast, the historian C. L. R. James has described the seafarers in Herman Melville's *Moby-Dick* as "a world-federation of modern industrial workers [who] owe allegiance to no nationality [...nor] to anybody or anything except the work they have to do and the relations with one another on which that work depends."[22] But even James, in an earlier passage, laments the "racial doctrine" that impoverishes political imagination.[23]

Life and work in our logistical age echo extraordinary changes in the nature of global capital: space annihilated by time, more technology and technical knowledge, faster loading, unloading and turnaround time, if not necessarily faster journeys.[24] Whatever shape contemporary exploitative practices take, there are continuities with life and work at sea centuries ago: common forms of uncertainty, precarity, and exploitation; shifting understandings of "race" but comparable configurations of racialisation; and similar subtle and magical transformations of the body to accommodate the chronotope of maritime commerce.

Like other forms of labour, seafaring is deeply enfleshed and experienced in the body. Paraphrasing the queer disability activist Edward Ndopu, the anthropologist Vanessa Agard-Jones reminds us that "If the body is a product of social relations, [...] it is also a product of assemblage: a melding together of blood, bones, flesh, chemicals, chromosomes, cells, and spirit."[25] The seafarer's body adapts to water, sea, and sky; it carries the memory of life and labour in the wheelhouse, the engine room, or the deck. The body is protected by masks when scraping rust from the skin of the vessel; by hardhats on deck; and by steel-toed boots and heavy canvas overalls in the hot bowels of the ship. Seafarers are injured more frequently than those who work on land. Because freighters never carry doctors and evacuations are exceedingly rare, injured seafarers have to be treated on board by first aid officers. Eyes, hands, and feet are most vulnerable to the maw of machinery, the burn of chemicals, and the scalding of lubricating oils. Skin can blister or burn in the engine room and bones can be crushed and broken by enginery and snapping ropes. A study commissioned by the International Transport Workers' Federation shows that twelve percent of seafarers reported injuries, a fifth of which resulted in hospitalisation, and thirteen percent in work restrictions or disability.[26] Work at sea shapes the body whether or not the seafarers wish it. The body adjusts to the feelings and motions of being asea. Organs, hearts, and legs change.

For a few brief decades in the twentieth century, seafarers in the global north could depend on professional positions within shipping companies. Even then, an account of seafaring described ships as "sweatshops at sea."[27] By the 1970s, work aboard ships had become unmoored from full-time employment for the vast majority of the world's seafarers. Legal and illegal recruiters, generous or unscrupulous, began connecting desperate jobseekers to ravenous shipping companies via short-term contracts. Blacklists and something approaching debt bondage could shape the terms of the seafarers' employment. In a way, seafaring on these precarious contracts was a precursor to gig work in other industries, but especially in logistics and delivery businesses.[28] Seafaring anticipated all these technologies of discipline and exploitation, from bonded work, to factory sweatshops, to precarious gig labour. Seafarers' bodies bore the marks.

FLESH

Nowhere does work seem more dangerous than when seafarers on cargo vessels get caught in naval crossfires. Four years into the calamitous war between Iran and Iraq (1980–88), the former had begun to take the upper hand on land. Revolutionary Iran was considered a pariah by the North Atlantic powers while – surreptitiously or overtly – they lavished arms and intelligence upon Iraq. France supplied Baghdad with Super Étendard fighter aircraft and Exocet missiles to replenish Iraq's diminished firepower. Using this new armoury, Iraq took a new tack in its maritime posture, and began to attack tankers carrying Iranian oil at the Kharg and Larak oil terminals in the Persian Gulf, both to enfeeble the Iranian economy and to draw in outside powers. It succeeded in the latter. In the ensuing four years, the two countries attacked 411 ships, sixty percent of them tankers. MT *Seawise Giant* was damaged so badly in an Iraqi aerial bombardment that it had to be rebuilt from scratch in a shipyard in Singapore. To forestall attacks by Iran on its allies, the US lent its flag to Kuwaiti ships, and convoys of European and American warships escorted tankers out of the Gulf. The last tanker to be attacked, Norwegian-flagged Merchant Tanker *Berge Lord*, was hit by an Iranian patrol gunboat only two weeks before the two countries agreed to a ceasefire in August 1988.

In early 1985, as the tanker attacks reached their first crescendo, Reverend Ernie Arnold of the Mission to Seamen (later changed to Seafarers) in Dubai recounted an incident in his monthly report to the organisation's headquarters in London. A "badly damaged firefighting support vessel," believed to have been hit by Iranian missiles, was towed into the Dubai Drydocks. The ship was, Arnold said,

> [...] a mass of twisted and burnt wreckage. The survivors and the two dead seamen had been taken off in Bahrain and the captain was missing, presumed blown into the water.
>
> Last Tuesday (nine days after the boat came to dock in Dubai), I received a telephone call from the local manager of Swires (sic) International who wanted to see me urgently.[29] He then told me that someone had discovered some signs of human remains in the shattered wheelhouse and that the captain was somewhere there. In the local situation there could be no help from either the police or hospitals to help in the recovery so it fell upon the local manager to do the job. It ended up with him, a doctor and myself spending about three and half hours on Thursday gathering together what we were able to find by sorting through the ashes and other remains in the wheelhouse. That evening just before the light failed, we took about three-quarters of a bin-liner out to the Indian burning place and placed it upon a funeral pyre and read the committal service.[30]

The Mission's harrowing wartime reports have a baroque quality. They include stories of death, injury, fire, destruction, and fear. Many of the unnamed mariners they record seem to have been from the global South; the captain described by Reverend Arnold was presumably Hindu, i.e. from South Asia. Mission chaplains in Dubai and Bahrain detail mariners' constant terror at sailing into the Gulf on their tankers, supply vessels, and cargo ships. They worry about being attacked, about never being able to leave. And though more than 400 civilian seafarers were killed in these attacks, these mariners – fearful, terrorised, bloodied, torn-apart – are absent from most accounts of the war's events,

The Tanker Wars; photograph courtesy of Norbert Schiller

and from much writing about shipping, trade, and commerce during the war.

The casualties of the Tanker War tally ships, not seafarers. We know the names of the vessels, which flags they flew, and what cargo they carried. We do not know the names or nationalities of the injured or the dead. Nevertheless the body of the ship's captain – his burnt flesh and shattered bones – is viscerally present in Reverend Arnold's report. There is no oblivion in this story, just lacerated bodies and harrowed souls.

MUSCLES

In his study of British colonialism in the Indian Ocean, maritime historian Frank Broeze calls seafarers the "muscles of empire," without whose stamina and strength the "British merchant fleet in the Indian Ocean would have ground to a complete standstill."[31] Broeze's label is no metaphor: the work of seaborne trade depends in fundamental ways on the embodied, visceral, muscled work of the seafarers. But these muscles are at once powerful and vulnerable. They are exposed to illness and the elements, but are also inseparable from potentially mutinous, intransigent subjectivities. For these reasons – vulnerability, mutiny – maritime technologies have been from the very start intended not just to extend but to replace human muscles at sea.

In the age of sail, ships were overmanned for this dual reason.[32] Steamships then required even larger numbers of workers: hauling and shovelling coal and maintaining the engine required people working below deck around the clock. But much changed with mechanisation and the invention of tank ships at the turn of the twentieth century. Tankers stored fuel and other liquids in large voids in the hold of the ship, rather than in barrels or other containers; their use of fuel oil and automated piping systems removed the necessity of multiple shifts of firemen and trimmers. The arrival of multi-modal freight containers and container ships in the mid-twentieth century accelerated this trend towards fewer bodies on board. These mechanised vessels required fewer seafarers to operate them, and fewer stevedores on the docks to load and unload them.

When port authorities in London and Liverpool attempted to break workers' strikes in the 1960s, they brought in consultants

The anchor chain on *Corte Real*

from the global management firm McKinsey & Company to advise them. McKinsey suggested that "expensive labour can be replaced with cheaper capital equipment."[33] Worker intransigence, unmentioned in the report, was its pervasive subtext; automation was a solution to unruly dockers. This, McKinsey reasoned, was a strategy proven in another maritime transport sector, namely the tanker industry. "The distribution of petroleum products is highly automated," they wrote. "The combined use of supertankers, high-speed automatic pumps, and large-capacity inland pipelines result in an extremely efficient integrated transport system."[34] McKinsey boasted that replacing humans with machines had saved the oil industry up to thirty percent of its earnings – a staggering amount of money.[35] Flags of convenience further reduced minimum manning requirements.

Aboard ship, work for seafarers – many previously accustomed to hard agricultural or factory work – can often be backbreaking, constant, and tedious. Though many tasks

Looking into the hold of the ship

have been mechanised on today's containerships, tankers, ro/ros (roll-on/roll-off vehicle carriers), and bulk vessels (which carry unboxed cargo such as coal, ore, or grain), labour is still repetitive and stressful, especially when arriving at or departing ports. Often seafarers are required to do tasks on arrival that should, under more labour-friendly bargaining agreements, have been performed by dockers. Where labour is cheaper than automated equipment, lithe and capable seafarers still perform tasks that would otherwise be assigned to machines.

Marcus Rediker describes the labour aboard sailing ships as involving perpetual shadow work including "overhauling the rigging, coiling ropes, repairing and oiling gear, changing and mending sail canvas, tarring ropes, cleaning the guns, painting, swabbing and holystoning of the deck, and check-ing the cargo."[36] Similarly, Alan Villiers's account of Kuwaiti sailing *dhows* in the early twentieth century depicts endless tasks to keep the vessel sea-worthy, much of it involving the maintenance of the sails, the ropes, and the wooden hull of the boat.

Steamships demanded an endless litany of drudgery and danger, in particular for firemen and stokers hauling coal to the engine-room and heaving it into the engines. These workers were called, in the anglophone world, "the black gang" because their skins were coated perpetually in coal soot. In the Indian Ocean, men of "the black gang" were recruited from Aden, Somalia, or Sylhet; in the racial determinism of the time, they were thought better able to stand the heat in the stokehold. Interestingly, though such racist presumptions presented a rationale for hiring South Asian, Adeni, and Somali seafarers to work in the hellish bowels of the ship, colonial categories also had a hand in assigning seafarers to this thankless job. A census of seafarers just before the First World War shows that "22% of all Irish seafarers were firemen or trimmers compared with 14.6% of English and Welsh" sailors.[37] The "black gang" embodied colonial and racial categories even as it described properties of working with coal.

In his account of *lascar* labour, scholar G. Balachandran describes work in the stokehold in horrifying detail:

> Scalding hot workplaces where temperatures could exceed 60°C especially in the Red Sea, engine rooms have with good reason been described as 'hell holes' [...] Burning, scalding, and heat asphyxiation were the common lot of engine-room crews particularly in the Mediterranean and in tropical waters where firemen had often to douse themselves with buckets of water before opening the furnace door and after shutting it. [...] To the dangers of accidental death in the engine room one may add that of being crushed by sudden shifts in heaving masses of coal. [...] Coal bunkers could be at a considerable distance from the

Manually unlatching boxes from one another atop the containers on *Corte Real*

engine room on large vessels, with depleting coal stocks further increasing the distance as the voyage progressed. Trimmers developed permanent bruises (or 'badges') on the side of their shoulders from knocking repeatedly against narrow, heaving walkways whilst attempting to balance themselves against the roll and pitch of the vessel as they wheeled coal from the bunkers to the engine room.[38]

The trimmers, the lowest paid men on the ship, literally bore the marks of their work on their bodies as badges.

The engine rooms of container ships are still hellishly hot places today, and like coal bunkers, they are below sea level: subsurface. The engine room is, however, not the sooty coal storage of the days of steam. It is instead gleaming and metallic; heat reflects from instruments and machinery. Pipes and machines in the room where lubricating oil is heated before it greases the engine are wrapped in heat proof foil;

Engine cylinders in *Corte Real*

still the place scorches, especially as we pass through the steamy heat of the Red Sea. The drive shaft is a meter wide, and the pistons needed to revolve it are larger than grown men. It is hard to imagine such sophisticated machinery being repaired by hand. The dozen-or-so men who worked in the engine rooms of *Corte Real* and *Callisto* were, however, true artisans; they did miraculous repair work on broken parts of this complex and massive engine with ordinary screwdrivers, wrenches, and other tools that might be found in a tool-shed. In the hot space below deck, engineers and electricians ensured that the thrum of the engine was reassuringly constant. They maintained the machinery's electrical and mechanical innards. It is hard to imagine how a fully-automated ship – the desiderata of shipping moguls – would repair itself in this way.

Above deck, the crew fought a constant war to contain the rust and salt gnawing at the walls and bodies of the ship. Sanding, washing, and painting never ended: the ships were so vast that by the time one part of the ship was cleaned and repainted,

Painting the hull

another would need work. Deck sailors also had to keep regular watch over refrigerated containers to ensure that their temperatures were suitably cold. Refrigerated boxes – 'reefers' in maritime parlance – often carry chemicals; unexpected changes in temperature can be hazardous.

In the wheel room, the tedium involved endless paperwork. The first mate cut and pasted new instructions on the Admiralty charts. Messages from port control or various maritime security organisations were received and filed. When arriving and leaving ports or crossing canals, bureaucratic documents were completed, signed and submitted. Hefty binders of certificates and paperwork jostled for space on the shelves of the wheel room with instruction manuals and shipping directories. At sea, officers in the wheel room mostly kept watch, looking to long distances and various screens to ensure the ship never got too close to another ship, or to subsea reefs, ridges, islands, and submerged volcanoes.

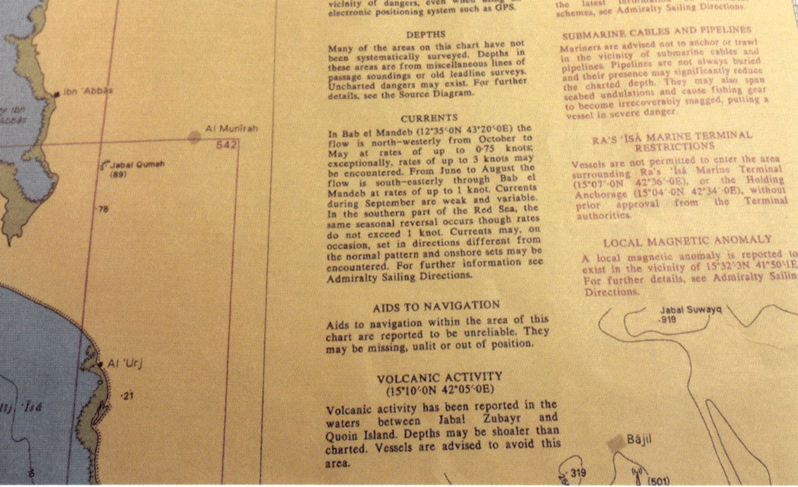

Admiralty chart showing undersea perils

The nature of sociability on ships has changed. Aboard the sailing ships of old, seafarers told stories to entertain one another. In *Outlaws of the Atlantic*, Rediker describes the meaning and genesis of the sailor's yarn. When the ship was at sea or in the doldrums, the captain created "make-work" to keep the sailors occupied and disciplined. Repairing the ropes was one such make work:

> As sailors sat together picking apart the yarn of their ropes, someone would spin a yarn for a bored, unhappy, unwilling, ready-made audience of common labor. The yarn, then, was in several ways a spoken-word equivalent of the work song. One of its purposes was to entertain, to help to overcome drudgery, to make the time pass, to transport both speaker and listener to a different, better place. It was, in short, born of alienation at work aboard the ship, which proved to be a nursery of narrative talent.[39]

The language of seafaring developed in these sessions of ravelling and unravelling, an international argot whose words came from the life of work. Like much else at sea, it developed as a way of negotiating quotidian life aboard a ship with others who do not speak one's language.

For Herman Melville's whaler, the "gam" is "a social meeting of two (or more) whaleships, generally on a cruising-ground; when, after exchanging hails, they exchange visits by boats' crews; the two captains remaining, for the time, on board of one ship, and the two chief mates on the other."[40] In his 1938 voyage on a sailing *dhow*, Alan Villiers describes something similar. Not only did the captain visit other ships while the two pulled abreast at sea, but also other seafarers aboard did as well:

> Life on board [companion ships] went on very much as it did aboard ourselves; at least once a day we somehow managed to be close enough to one of them for a yarn, during which the masters, mates, and crews of the vessels conversed quietly across the few yards of intervening sea and now and again borrowed things from one another. We heard their muezzins, morning after morning, give the call to prayer: there used to be a kind of a race between three announcers.[41]

This form of sociability has disappeared on ocean-going ships as ship sizes have ballooned; one cannot imagine them coming abreast at sea without an accident. The seafarers of today socialise in other ways. In the ships I travelled on, European officers would gather in the officers' sitting room, drink beers, and watch football games, films, and television shows in Serbo-Croat. I was never invited to their social events, perhaps because they couldn't (and shouldn't have to) bother with speaking English to make a nosy passenger feel welcome. On my first journey, I was invited to karaoke sessions in the crew's recreation room; Filipino crew members often requested a karaoke machine as part of their contracts. The gatherings were warm and noisy, and seafarers were friendly and curious about what I was doing on their ship.

On the second journey, CMA CGM had extended satellite internet connections to seafarers. Before that, only the captain had access to satellite wavelengths, and would gather emails and communications from the officers and crew and send them in bulk at night. Once the crew had access to internet, the socialising around karaoke pop songs receded and they often stayed in their cabins conversing with family members, cruising the web, or watching porn, things they had previously only been able to do at port, and even then at great expense. The claustrophobic intimacy of the ship, its weeks long isolation at sea, and the non-existent line between place of work and place of leisure, meant that sometimes the seafarers wanted to escape one another by retiring to their (shared) cabins to speak to their families. But it seemed to me that although the men could connect to partners, children, and families more easily – most beamed while showing me photographs of their children captured on Skype – their lives aboard the ship had become more solitary and less sociable.[42] Only when the captain planned a barbecue on deck – the crew told me it was thrown in my honour, as it rarely happened without a passenger onboard – did seafarers, or at least those not on shift, get to socialise across crew/officer boundaries. Grilling and eating arrangements were nonetheless determined by language and nationality differences. Moments of leisure were embedded in the rituals of a workplace – albeit a mobile one – and were shaped by the diktats and constraints of labour regimes and hierarchies.

In the interstices of work and relations between officers and crew, bodies were shaped by seafarers themselves. Most large freighters contain gymnasia with cardiovascular and table-tennis equipment and weightlifting paraphernalia. On both ships, European and Filipino seafarers had different

The gym aboard *Callisto*

ways of crafting their bodies according to masculine ideals. Every Croatian officer was brawny and spent hours pumping iron. Even the woman cadet on the first ship had a muscular upper body and prided herself in having physical strength to lift above the expected level for her weight. By contrast, the Filipino seafarers preferred cardio exercise or playing table-tennis; their idea of strength was not necessarily related to being muscular or top-heavy. Kale Fajardo, an ethnographer of Filipino seafarers, has described how "sailors and tomboys" craft alternate forms of masculinity that challenge the neoliberal modality exported worldwide through practices of body-making.[43] Seafarers' crafting of their physical selves challenges the neoliberal attitudes reinforced by the Philippine state through heteronormative and masculinist structuring of maritime education.

The shipboard life of flesh – sexual relations aboard – was not my story to tell.

No narrative of shipboard work would be complete without a discussion of food and the rituals surrounding meals. Aboard ships, "food has a social meaning which is both more significant and different from what is the case in families ashore."[44] Disparate food is provided to crews and officers, and the rituals of eating also differ. On both journeys, the officers' mess was separated from the crew's dining room by the kitchen. The officers' tables were set hierarchically, with the captain at the head, his first mates and the chief engineer on either side, and so on down the table. The seat of the Filipino second mate was always left unoccupied – he ate in the crew's mess. Filipino officers joined Filipinos on the crew, rather than eat with their fellow officers. Passengers ate at separate tables in the officers' mess and were served the same food. Eating was a ritual strictly and punctually maintained. Meals were served at 6am, noon, and 6pm. If you missed a meal, you sated your hunger with hardboiled eggs, apples, or Danish cookies (served in round tins that people throughout the Middle East use as their sewing kit boxes).

Our Filipino chef had worked at several expensive hotels in Dubai and aboard a cruise ship. He often prepared standard meat-and-two-veg dishes for the Croatian officers. After a few days of boring meals, I timidly asked if I could be served what the Filipino crew was eating. The chef and messboy at first demurred, telling me that Filipino stews were made with meat left over from the officers' meal the previous day. I finally prevailed and for the rest of my journey was fed delicious *adobos*, curries, and other Southeast Asian stews that were, though sometimes unfamiliar, more flavourful than the dull European menu. There, in the basic ingredients of the food, in the way they were made, in how they were eaten, the global social order came into view. The crew's meals were made with untouched leftovers of officers' meals, and hierarchies aboard the ship were trumped by national attachments, with the Filipino officers eating with their countrymen, rather than with their officer cohort.

Food is also something more banal and non-negotiable: you need food to live. Hunger ships sail through maritime fiction with alarming frequency; the most harrowing twentieth century rendition occurs in B. Traven's *The Death Ship*. Traven is, perhaps, the pseudonym of a German anarchist who also wrote *The Treasures of Sierra Madre* and swashbuckling novels about the Mexican revolution. *The Death Ship* tells the story of a sailor forced to steam on ships days away from sinking – or, rather, being sunk, often for insurance money. Aboard various incongruously named rattletraps, sailors are "always hungry because a shipping company cannot compete with the freight rates of other companies if the sailors get food fit for human beings."[45]

Historical accounts of shipboard life on the Indian Ocean abound with the inadequate victualling of ships manned by *lascars*, and intransigent owners refusing to increase rations by even the smallest margins.[46] Where any improvements in rations were wrested from shipping firms by European unions, *lascars* were excluded from the meagre increases. *Lascars* were often so badly fed, so badly clothed, and so badly accommodated, that they had disproportionate rates of disease and death onboard.[47] On *dhows*, seafarers provided their own meals and could starve if bad weather or inadequate fuel meant being stranded at sea. On all ships, seafarers often resorted to fishing to supplement their meals. It was known as early as the eighteenth century that a paucity of certain nutrients could lead to beriberi or scurvy. Yet shipping firms were too miserly to provide even dried citrus to their seafarers to forestall illness.[48] Shortage of food was the most frequent cause of mutiny aboard ocean-going ships in the Indian Ocean.[49] And, of course, inequalities aboard ships meant that the "black gang" in the belly of the ship suffered more because "the heat of the engine room reduced food to a pulpy, tasteless mess; vegetables were rarely fresh; butter became liquid and, although stored in open-air lockers on the poop deck for coolness, potatoes were no better."[50]

Hunger ships perennially betide the shores of the Arabian Peninsula because of the sparsity of labour safeguards, the dense interweaving of networks of capital accumulation with the ruling courts, chronic criminalisation of unions, and limitation of other labour or seafarer rights groups. Hunger ships are often owned by shipping companies based on the Peninsula and abandoned on the shores of the United Arab Emirates, their seafarers left to their own fate.

In the decade after the gradual nationalisation of oil on the Peninsula, many oil-producing countries there invested a great portion of their oil wealth in building up fleets – both owned by the state and by firms with ties to those in power – to transport oil, dry cargo, and eventually containers. This proliferation of ships flagged to Arab states was not accompanied by robust registration or regulation systems. The archives and databases of the International Transport Workers' Federation and the International Labour Organisation are therefore rife with accounts of ships in that period effectively operating as if under the least-regulated flags of convenience. Minimum wages were not stipulated, unions were not allowed to operate, strikes were (at least in Saudi Arabia) prosecuted as political conspiracies. Insofar as most ships flagged to Gulf countries were manned by foreign seafarers, the workers did not have the protection of citizenship either. In the internal correspondences of the ITF, all sides effectively consider Saudi, Emirati, or Kuwaiti ships as *de facto* flags of convenience, even if they were not formally designated as such. The archive contains a letter

from a Sri Lankan captain who describes how he resigned
from the Saudi Europe Lines because he did not agree to the:

> illegal suggestions and operation methods carried out
> by the Saudi Arabian Owner (sic) and his Greek staff.
> [...] It would not be too long when ships of the Arab
> world will have the largest fleets, due to concessions etc
> enjoyed by Arab owners, bunkers at a very nominal
> price (which is of course misused), permission to oper-
> ate ships of 20 years and more, whereas restrictions been
> brought to bar ships of over 20 years of other nations
> entering Arab ports, low wages and poor working con-
> ditions. [...] I wish to state without hesitation as I have
> been Master of Arab flag ships and my experience is that
> ships of these flags are the biggest violators of all inter-
> national regulations.[51]

In extreme cases these conditions have transformed ships
flagged to the states of the Arabian Peninsula into sites of starvation and despair. In my 2020 book *Sinews of War and Trade*,
I tell the story of a Saudi-owned and -flagged ship which, on
a 1981 journey from Saudi Arabia to Europe, progressively
became a hunger ship. The story is told by the Filipino seafarers themselves, second or third hand, and anonymously. I
found the narrative in the appendix of an MA dissertation
from the 1980s. The appendix itself was a photocopied version
of a 1981 typewritten report by the Human Rights Project
of the US-based Seamen's Church Institute. The report had
excerpted yet another document dated 2 June 1981, produced
by the Commission for Filipino Migrant Workers, which quoted
an unnamed Filipino seafarer:

> For ten months the *Saudi Independence* has been sailing
> as a hunger ship. Despite repeated requests of the crew
> for adequate and varied food, the shipowner has refused
> the crew's request. The ship sailed from port to port,
> from Jeddah to Piraeus, to Antwerp, to Bremen, etc.
> The shipowner always promised that the food supply

would be replenished in the next port. But already after some weeks, the captain at that time was dismissed because of his efforts to improve the food situation on board the ship. His place was taken by a Filipino captain. Soon after he was sent back to the Philippines for the same reason. Later, during the second trip to Europe, the Filipino first mate was also dismissed after making a complaint over radio on the lack of food. Finally, the radio officer was also dismissed. The need became so great that the crew had to improvise making a fish net to try to catch fish and supplement their meager rations of food with fish. At the same time, the shipowner very shamelessly sent a telegram ordering more savings and limitations on food supply. There was in fact no food supply.[52]

When the ship arrived in Rotterdam, the Filipino seafarers revolted and downed tools. The shipping company, Jeddah-based Orri Navigation in turn brought a suit to force them back to work. In the court proceedings that followed, the Dutch justice system ruled for Orri Navigation. The legal case notes induce rage, not least because of their phlegmatic matter- of-factness. The Dutch court ruled that the seafarers' contracts bound them to Philippine law, because more favourable Dutch laws could set a precedent of further strikes and labour action by seafarers from places whose legal systems denied them these basic rights. The court also decided against the legality of the strikes on the procedurally pedantic basis that the ITF which had represented them was not "a real union" and that two months' notice was not given for the strike. Finally, the strikes were dismissed because "the strikers originally complained only about the food. Their complaints about the wages were made only after contact with the ITF."[53] The seafarers were fired without any wages after having spent weeks hungry onboard.[54]

The monthly reports of the Mission to Seafarers are filled with stories of such hunger ships at the ports of the United Arab Emirates, Bahrain, Kuwait, and Oman. One report describes the plight of a Croatian crew aboard a ship at Jabal Ali in 1993:

> After about my third visit the situation seemed to be deteriorating and I realised they had no fresh food to speak of, so drove down to Dubai to get vegetables and frozen chicken for them and to speak to the agent, who in fact said he hadn't realised the situation was so bad. On returning I realised they had even run out of water, but at that moment a lorry arrived to supply them with some. Two days later their diesel finished but I was able to verify with the agent that that day they were getting 200 dollars for food and oil would be supplied. I took them this good news, and some more food, but the ship was quite unpleasant as they didn't even have AC anyway and for a day or so they had had no pumped water or electricity. One man was just getting over a very bad case of prickly heat.[55]

A more recent case is equally shocking. An Emirati firm, Elite Way Marine Services, seems to have a history of abandoning ships. In 2017, Elite Way's MV *Azraqmoiah* carried building materials between Iraq and the UAE. When the carriage contract ended, the ship's owners abandoned the ship, with crew members – eight Indian, one Sudanese, and one Tanzanian – left at sea in an anchorage near the Emirati coast, with £260,000 in unpaid wages and without food or fuel. Nor did the crew have any way to repatriate to India, Sudan or Tanzania. To stave off starvation, Mission to Seafarers and the Indian consulate provided them with paltry rations of rice and dal, and the seafarers remained onboard to maintain the ship and to guarantee that they would eventually be paid their wage (unpaid wages can be paid out of the sale of an abandoned ship). After nearly eighteen months at sea, the seafarers were finally repatriated with forty percent of their wages after the ship was sold by Emirati authorities.[56]

Elite Way Marine Services also owned MV *Tamim Aldar* – another even more egregious case of abandonment. Stranded by its owners between 2016 and 2019, Tamim Aldar originally had thirty-six crew members, of whom two Indian and two Eritrean seafarers remain stranded at sea twenty-five nautical miles off the coast of the UAE. The four remaining seafarers had no fuel, little water, and scant food, and lived in the blacked-out ship. When they attempted to swim ashore, the Emirati coast guard forced them back, and threatened incarceration.[57] Their forcible return to the unsafe ship generated enough controversy for the Emirati authorities to allow the men to come to shore, but Elite Way refused to pay their wages in full, offering only fifty percent of what they were owed. The men remained in the UAE, stubbornly demanding their wages and transportation back to their homes.[58] International publicity around the case, in the end, forced Elite Way to pay the four remaining seafarers eighty percent of their pay. They were eventually repatriated to their homes, courtesy of the ITF.

In a monthly port report, a representative of Mission to Seafarers described the emirate of Ajman as an anchorage "where you can find some very dodgy ships."[59] According to the International Maritime Organisation's senior legal counsel, the UAE is one of the worst culprits for sheltering abandoned ships.[60] But ships are not abandoned randomly; they are rarely left close to Sharjah or Dubai or Abu Dhabi, where the ports are crucial to the operation of the economy; as richer emirates, they have enforcement abilities denied to Ajman. Like all other countries of the Arabian Peninsula, UAE refuses to become a signatory to the Maritime Labour Convention which would insure pay for unpaid wages of seafarers. Nor are abandoned seafarers a random sample of all seafarers. Abandonment disproportionately affects seafarers of Indian and African origin, even as the latter are a smaller percentage of all seafarers on ships. The global colour line runs even through cases of shipboard starvation. Unimaginably, the horrific conditions that B. Traven sketched a hundred years ago in *The Death Ship* prevail even today.

It is a paradox of shipboard life that loneliness reigns in a place where physical intimacy and proximity define everyday interactions of seafarers. Ravi Ahuja quotes Amitav Ghosh on how aboard ship, "every man's nose was inches away from a solid barrier: either the ceiling or an arse."[61] Ahuja goes on to describe how this "quotidian, painful experience of confinement and monotony" could sometimes lead to the suicide of *lascars*, or their falling into madness.[62] He retells the story of one sailor who burnt his own eyes with caustic because he had seen *Shaitan* (the devil) below deck. In his mid-twentieth-century ethnography of Afro-Iranian life on the Gulf coast of Iran, Gholam-Hossein Sa'edi writes of the wind (or spirit) that blew from the sea and possessed specific categories of inhabitants of the shores:

> Fishermen, seamen, and the women who work on the palms are more than anyone else prone to catch these spirits. Those who have good and comfortable lives, such as tradesmen and the captains of larger ships, are never subject to these winds [...] Seafarers who always live at sea are often afflicted by coastal winds. Nary a sailor can be found who has travelled to Africa or India once or twice and has returned without these winds.

He describes how seafarers pressed into service, often under threat of violence, surrendered to possession after long periods at sea. The cure for such possession was to remain sequestered at home for a time, implicitly providing a break from conditions of forced labour.[64]

If monotony and exhaustion define life on ships, port visits provide a respite. Taverns, coffeeshops or teahouses, marketplaces, tattoo parlours, and brothels historically served the sea-farers in their periods of liberty in port. As ships have become larger and ports have moved further out of city centres, periods of liberty have diminished in length and ebullience. "Productivity" and "turnaround efficiency" now dictate how long seafarers can remain "idle." In the early twentieth century, worries about sexually transmitted illnesses and drunkenness were precipitated more by a concern about productivity rather than the health of the seafarers themselves. The moralist discourse of protection against disease and intemperance was from the start used as a justification for ever-shortening shore liberty.

A 1956 *New York Times* report on work on mega-tankers commented, "At sea, the tanker man lives pretty much as does the freighter man, except that there is more of it. For the big tanker spends little time in port. She is in this afternoon, to start spilling out her cargo, and after eighteen hours is off again tomorrow, for a month or two."[65] The report casts a prudish eye on this life of solitude: "One advantage in his way of life may be the money – he is usually somewhat better paid than the freighter man and he can save a lot, for there isn't much to spend his wages on. The oil loading ports in faraway lands are isolated, not near the white lights. A man can use up most of his day in port getting to and from the nearest hot spot."[66]

Even when port visits were longer they did not repair feelings of loneliness and homesickness. Nikos Kavvadias, former mariner and Greek bard of the seas, describes his liberty in Marseille:

> I ripped up a lot of paper writing this;
> it's made me dizzy being in Marseille.
> A moment doesn't pass, believe me dear,
> without my thinking, even here, of you.
>
> Northern sailors and stokers from the South
> sit with girls they pay for on their knees;

the piano plays itself, a young girl whistles
a rough approximation of a well-known tune.

As I came out drunk from the bar "Tartan"
my body seemed so worthless and small
and I felt you smiling there beside me
with that strangely bitter laugh you have.[67]

In another poem, Kavvadias writes of *Fata Morgana*, or ship-board hallucinations that afflict seafarers on vessels that he describes as "the holy rust that bears us,/ feeds us, is fed by us, and finally kills us."[68] Narcotics or intoxicants briefly assuage the sense of loss. I have not read a single maritime story in any language where seafarers do not consume something – tobacco in whatever form, herbs, leaves, white powders, fermented drinks – to allow them to persevere. A whole genre of pulp novels burgeoned in the 1950s and 60s recording the life of tankermen at sea with all its debauchery. Robert Mirvish, the master of the genre in English, wrote stories about "pleasure, drink and women in foreign ports" replete with orientalist portraits of natives.

A 2019 survey of seafarer mental health commissioned by the International Transport Workers' Federation found that "25% of seafarers completing a patient health questionnaire had scores suggesting depression (significantly higher than other working and general populations)." Even more alarmingly, the researchers reported that "20% of seafarers surveyed had suicidal ideation, either several days (12.5%), more than half the days (5%) or nearly every day (2%) over the two weeks prior to taking the survey."[69]

Fajardo's ethnography of contemporary Filipino seafarers briefly notes suicide as a form of resistance to or escape from the difficulties of work aboard ships.[70] Suicides and unexplained drownings and disappearances at sea continue to afflict seafarers – especially among deck ratings, stewards, and cooks – at triple the rate of landbound populations.[71] A 2010 occupational health study on British seafarers pointed out that seafarers cited work problems, depression, hallucinations, and marital or relationship problems as the most frequent reasons for self-harm.[72] The authors of the study let it pass without comment that the rate of suicides seemed to spike around 1970, at the very moment when open or international registries proliferated and hard-won labour regulations were eroded. The same study showed that *lascars* committed suicide at a higher rate than non-*lascars*. The global colour line was foundational to rates of self-harm, but also to the paucity of knowledge about suicides among seafarers from the global South. The authors of the 2019 ITF mental health report phlegmatically state that Filipino and Eastern European workers were most likely to be exposed to "workplace violence" and this was the most significant determinant of depression and anxiety.[73]

Rates of suicide increased during the COVID-19 pandemic as seafarers remained aboard their ships after the end of their contracts, sometimes for months on end, in fear of an unknown disease, and prevented from taking shore leave or returning home. The executive director of a seafarer welfare charity claimed in 2021 that both suicides and calls to suicide hotlines operated by the charity had "roughly doubled" since March 2020.[74] But like much else that has to do with workers, actual and comprehensive statistics about such self-harm was sparse, and the casualty numbers unspecified. An officer interviewed by a researcher during the pandemic laid out eloquently many factors that caused distress among the seafarers:

> My company is requiring all officers to remain on board
> for a minimum of 105 days, which far exceeds our regular
> 75, which I had made plans for and scheduled out the
> remainder of my year accordingly. It is hard for us to get
> stores. I find it ridiculous that we are not allowed to do
> crew change "for our own safety" however, we are forced

Shipwreck on the shore of Damietta, Egypt

to do cargo operations and interact with shore personnel who have been exposed to the outside world and potentially are ill in order to make them money. Then it's totally fine. All in all myself and the entire crew is upset, and very angry. And above all, what it has really shown me is that the company and even our own unions couldn't care less about us as people and only care about our earning potential for them. They'll lie straight to your face about it too. The entire process has been exhausting and very disappointing. Motivation is at low all-time, everyone feels as trapped as we are, and I've never wished I had a less "essential" job until now.[75]

With occasional access to the internet, other seafarers posted desperate pleas to be released from their confinement at sea. Uncounted numbers drowned themselves in the vastness of the sea that kept them captive.

FORKING PATHS

In 2020, just before the pandemic lockdown, I gave a public talk in Rotterdam at an event organised by Erasmus University. Rotterdam is the capital of shipping in Europe and I was a little nervous when a few Dutch seafarers showed up to the talk. They were union men; all had served on ships, some still worked on shore. Two worked for the Dutch equivalent of the Mission to Seafarers and one of them interviewed me for their onboard magazine. They sat somewhere in the middle of the arena-like seating. Their faces were unreadable as I gave the talk. One came up to me and exchanged requisite niceties afterwards. Anxious about my pretentions to expertise about seafaring in a talk with seafarers in the audience, I asked him what he thought. He demurred. So I pushed a bit more and he said to me, the story you tell is a story of suffering. We are *professionals*. We are good at what we do. Without us, none of this – and his arm swept the amphitheatre – would be here.

I have thought about that conversation often. When I have felt churlish, I have resorted to an all-too-simplistic response: "Well of course *he* would talk about profession-alism; *he* is not a relatively underpaid Filipino seafarer." Though ungenerous, this statement is not entirely wrong. European workers form an aristocracy of labour who have, over the course of many decades and even centuries, secured collective bargaining rights, good salaries, and the certainty of employment rather than precarious gig work aboard ships. This has sometimes come at the expense of workers from the global South. Histories of colonialism and empire have also shaped worker solidarity (or absence thereof) across the global colour line. European unions often take positions that benefit their membership along racialised lines and against the interests of other workers.[76] Dutch labour historian Marcel van der Linden has called this fissure *relational inequality*:

> Some workers are much better off than they were in the past, *because* other workers, who may also be better off than they were in the past, are vastly lagging behind in living standard and thence, are in absolute terms worse off than their Northern class brothers and sisters.[77]

The Dutch seafarers who spoke to me have indeed benefited from this relational inequality. They can take pride in their professionalism because their jobs are protected, well rewarded, and dignified. But this is not the whole story. The Dutch seafarer was right in other ways: perhaps less about concrete differences in the labour of Filipino and Dutch workers, or even their subjective experience of it, and more about the way I have chosen to construct the story of embodied labour aboard ships as a litany of suffering.

I have always been interested in how we choose to frame our intimate histories and memories. In my first book, I examined how Palestinians commemorated their past in two distinct periods: the first of national struggle during the heady times of Third-Worldism and global anticolonial revolt; the second after the end of the Cold War and with the ascendancy of discourses of humanitarianism and human rights. The forms that Palestinian narratives about their past took in these two periods were distinct, following an arc from epic heroism to tragic stories of trauma. I talked about how discourses of suffering functioned in the context of Palestinian mobilisation:

> Recognition of the suffering of ordinary people acknowledges their day-to-day sacrifices. It opens a space for the ordinary, often silent – though not always passive – majority who are usually overshadowed in the self-congratulatory glare of heroic [narratives]. [This

discourse of suffering] pulls back the curtain concealing the costs of mobilization for the dispossessed, and demands the world to see and to act. But in doing so, it shifts the onus of action onto the international community, makes suffering itself a virtue, and denies the possibility of agency, mobilization or collective action. The tragic discourse insists that the suffering of the past suffices as a necessary condition for acquiring rights in the future, and sometimes subverts the possibility of internal organization and mobilization.[78]

What do we gain if instead of lamenting trauma, we recognise what C. L. R. James has described as "the skill and the danger, the laboriousness and the physical and mental mobilization of human resources, the comradeship and the unity, the simplicity and the naturalness" of a collective of people at work?[79] After that conversation, I went through my notes, photographs, and diaries of my travels, and I discovered the subtle, unconscious ways I had chosen to tell the story the way I had – the way it had troubled the Dutch seamen.

How might I tell the story of seafarers at sea in two different ways at once? Jorge Luis Borges's gnomic short story "The Garden of Forking Paths" provides a clue. In this enigmatic tale, Borges writes about a Chinese spy visiting a British sinologist and former missionary named Stephen Albert, who is at once "priestlike" and "sailorlike." In an oddly intimate conversation between these two strangers, Albert unravels the mystery of a two-hundred-year-old labyrinthine Chinese novel, designed like a garden maze:

> The garden of forking paths was the chaotic novel; the phrase 'several futures (not all)' suggested to me the image of a forking in *time*, rather than in space. [...] In all fictions, each time a man meets diverse alternatives, he chooses one and eliminates the others; in the world of virtually impossible-to-disentangle Ts'ui Pen, the character

chooses – simultaneously – all of them. He *creates*, thereby, 'several futures,' several *times*, which themselves proliferate and fork. [...] Once in a while, the paths of that labyrinth converge: for example, you come to this house, but in one of the possible pasts you are my enemy, in another my friend.[80]

Setting aside Borges's gentle dig at the heterotopic occupations of Stephen Albert – a missionary, a sailor, and a colonial knowledge producer – this garden of forking paths offers the possibility of multiple futures thriving in the imaginative space of a human *not* choosing a path from among diverse alternatives, one that acknowledges the suffering *and* the skills; the striations and the comradeships.[81]

So onwards to another forking path in the story of seafarers at sea…

On both of my journeys from Marsaxlokk to Jabal Ali, when the ship reached the pirate waters of the Gulf of Aden, beyond the Bab al-Mandab and into the Indian Ocean, the ships went into alert mode. A range of safety measures were carried out. Openings on the lower decks were closed, and firehoses and other defensive measures were wheeled out. Our ships' freeboard – the distance from the deck to the sea – was too high for most pirates, who throw grappling hooks onto the deck and climb up on rope ladders. Russian and Israeli ships often carry armed guards, but not these ships, which depended on the alertness of the seafarers. Among the special measures were additional watchmen in the wheel room staring out to sea.

Mega-freighters today are equipped with devices that pick out other ships and objects at sea, including radar, Global Positioning System (GPS), Automatic Identification System (AIS), and other electronic equipment, supplemented with high-powered and heavy binoculars. The seafarers were able to look out to sea, keenly exceeding the abilities of sophisticated equipment: they could see further and better than I could with binoculars. They could distinguish a pod of dolphins from a *dhow* or a fast-moving skiff (the preferred pirate vessel). At first I didn't believe them, especially as earlier they had teased me that they could see skiffs speeding towards us at twenty-seven knots, twice the speed of our lumbering ship. But then they pointed to a far distance to something that did not show up on radar. In a few minutes, a pod of dolphins passed our ship. This transformation of their sensorium – eyes trained over decades to see further than logically possible – speaks to how bodies are shaped through repetition and practice.

An EUNAVFOR ship and helicopter in the Gulf of Aden

I was also struck by how the Filipino seafarers insisted on maintaining (and showing off) their knowledge of navigation by stars. Familiarity with the constellations – their seasonal revolutions, the directions they pointed – and the ability to read longitude and latitude using analogue instruments and navigational astronomy seemed like skills the Filipino officers (but not the crew) were taught in maritime academies. This was unlike younger European officers who had been trained to work only with digital navigation equipment. We had, during both of my trips, news of ships whose digital navigation instruments had failed or which had been spoofed for nefarious purposes, including one instance where such failure led to a US naval vessel colliding with another ship. In 2016, the US Navy had reinstated the requirement for naval officers to re-learn stellar navigation, only ten years after discontinuing it. Security concerns about the hacking of electronic navigation systems were behind this change in policy.[82]

Admiralty chart showing anti-piracy convoy route (parallel lines) through the Gulf of Aden

This equivocation between embodied knowledge and automation has been a constant thread of lamentation in maritime stories. Before the advent of machines that translated human knowledge into tools, observation of the stars and their motion was considered central to navigation, as were skilled predictions of ocean currents and a sense of how to read surfaces, marine protrusions, and coastal conditions. In the Indian Ocean, seafarers measured the altitude of a star above the horizon line using finger widths, and used the relative location of Polaris, the North Star, to determine latitude.[83] The coming of steam-power meant a proliferation of narratives about the loss of knowledge of winds and currents, and with metal-hulled ships, concerns about the demise of skills in building wooden ships. As new instrumentation emerges, worries about reading maps, understanding radio signals, and stellar navigation arise. Margaret Schotte, a historian of navigation science, has explained that "this shift from deck to desk, and from practice to theory, mirrors the broader transformations that characterized the Scientific Revolution of the sixteenth -and seventeenth- centuries, when many early modern Europeans began to view the natural world through a lens of rules and mathematics."[84]

Eugene O'Neill has lamented the arrival of steam ships in his maritime play *The Hairy Ape*:

> 'Twas them days a ship was part of the sea, and a man was part of a ship, and the sea joined all together and made it one. Is it one wid this you'd be, Yank – black smoke from the funnels smudging the sea, smudging the decks – the bloody engines pounding and throbbing and shaking – wid divil a sight of sun or a breath of clean air – choking our lungs wid coal dust – breaking our backs and hearts in the hell of the stokehole – feeding the bloody furnace – feeding our lives along wid the coal, I'm thinking – caged in by steel from a sight of the sky like bloody apes in the Zoo!

Nostalgia about the age of sail, however, belies the violence of those centuries while recognising the brutalisation of steamship labour. In another context, Raymond Williams describes nostalgia as resulting from changes to social relations and urban life, and as acknowledging of the fundamental "alteration of perception and relationship" under capitalism. "What was once close, absorbing, accepted, familiar, internally experienced experience becomes separate, indistinguishable, critical, changing, externally observed."[85] Seafarers' nostalgia for better working conditions in the past pulse with fear of losing their jobs to automation, of being burdened with unrealistic expectations of productivity, or the very real possibility of under-manning aboard ships. These fears sit alongside the pride seafarers have in their artisanal skills: as recently as July 2023, a Filipino seafarer who uses social media to record the lives, anxieties,

Reading the night sky in the Gulf of Oman

and joys of mariner cadets, wrote, "Steadfast and seaworthy as can be, we cadets are certainly ships embodied by our own human free will, ready to embark on the real word."[86] The cadets can not only read a map on paper, but they can locate themselves at sea using stars. Such expertise has commercial benefits that are enriched by the cosmological connections it establishes between seafarers and their environments. One of the officers who showed me his knowledge of the night sky also – in a rare moment of vulnerability – described how different the sky looked over his home village in the Philippines. The map of the stars wasn't just an industrial tool to him; it was also an embodied way in which he connected the institutional isolation of the ship to his home thousands of miles away.

Working at sea transforms bodies in other ways too. Before I saw so many seafarers in one place, at the seafarers' hotel in Marsaxlokk, I wasn't sure what I was expecting – but something rowdy, loud and salty. I imagined men (seafarers are still mostly men) who walked with a wide gait, expecting the solidity of land to give way to the sway of the sea. Of course, I was all wrong.

The seafarers working on the massive freighters of today do not have the gait of sailors of yore, who walked on land as they did on ship, keeping the body in balance. There was a time that the gait of sailors hinted not only at their walks aboard ships, but also their embodied masculinities. The great chronicler of capitalism, Honoré de Balzac, has described this wide stance:

> Sailors remain with legs apart, and always ready to bend or contract. Obliged to sway about on the deck with the swell of the waves, they are rendered incapable of walking straight on dry land. With all this sidestepping, they would do well to become diplomats.[87]

Ocean-going freighters, however, are flat bottomed and heavily laden; walking on deck is most of the time not that different than walking on land – unless the sea is stormy. We sailed through monsoon weather in the Indian Ocean; as ten-metre-high waves crashed against the hull, the ship did not list or feel unstable, though its body adjusted subtly to the motion of the sea with sways of its own. I joined the first ship I steamed on in Malta after it had come through the Bay of Biscay, where in heavy storms it had listed so far that all mobile furniture in the cabins had to be chained down and

CMA CGM *Balzac*, berthed in Khor Fakkan

loose items safely locked away. Stormy weather is where sea legs come into their own. A novice at sea will hold the wall as the ship lists from side or side or dives into the trough of a wave. More experienced seafarers adjust their gait and the tilt of their torsos to counterbalance the motion of the ship. The body becomes an extension of the ship, feeling its motions in bones and flesh.

The bodily mechanics of sea legs are also extraordinary in sketching the magical corporeal quality of adaptation and habituation. Recent scholars of physiology who have studied sea legs found that as days at sea go on, seafarers' walking stance widens, anchoring them more solidly to the ship. The same study demonstrates that the imperceptible sways that stabilise our bodies lessen when seafarers look to the horizon, reducing their seasickness, and that time at sea will synchronise the rhythms of their swaying to that of the horizon.[88] Another study found *mal de débarquement*, the sensation of disequilibrium as solid ground seems to pitch and roll under one's feet after coming ashore, is experienced disproportionately by menopausal women, leading the scientists to conclude that the sensation was not primarily precipitated by how seagoers hold themselves, but by the complex and little-understood movement of hormones through their bodies.[89]

Searching for "sea legs" in databases, one mostly finds metaphorical limbs, not flesh-and-bone or physiological ones. The expression has come to represent adaptation to one's environment, and confidence and prowess in what one does. The great Jamaican-American poet Claude McKay wrote two novels about seafarers and sex workers in Marseille. The first of these, *Banjo*, was published in his lifetime. It is the story of Black seamen from the United States, Africa, and the Caribbean arriving in Marseille in the interwar years, having jumped aboard ships as mariners or stowaways, and going as far away from home as possible. These men, especially those from the United States, enact their "dreams of vagabondage" to escape the constraints of sexual, racial, and social norms. This unmooring – or mass seagoing – is described as having "voted for freedom by exercising their sea legs" in a fifty-year-old essay published in *Negro History Bulletin*.[90]

In McKay's novel *Romance in Marseille*, posthumously published, the protagonist Lafala, a powerfully built West African seafarer, stows away on a ship crossing the Atlantic. He is discovered by the captain and detained in a freezing lavatory where his frostbitten feet become gangrenous. On arriving in New York, Lafala's legs are amputated. The loss is devastating. His legs were the tools of his trade and the delightful instruments of dancing:

> Legs like a quartette of players performing the passionate chamber music of life. Loud notes and soft, notes whispering like a warm breath, a long and noiseless

kiss, flutes and harps joined in enchanting adventures […] Legs of ebony, legs of copper, legs of ivory moving pell-mell in columns against his imagination. […] Feet that were accustomed to dig themselves into the native soil, into lovely heaps of leaves, and affectionate tufts of grass, were now introduced to luxuries of socks and shoes and beds of iron. […] Think not of age, of accident, the festering and mortification of youth and poisoned worms corroding through the firm young flesh to the sepulchral skeleton. His dancing legs would carry him over all.[91]

The American captain's cruelty cost Lafala's dancing and seagoing legs; but those legs, in their absence, become the engine of material transformation. He sues the shipping firm and unexpectedly receives compensation. Black newspapers celebrate his victory: "AFRICAN LEGS BRING ONE HUNDRED THOUSAND DOLLARS."[92] Sea legs become vessels of escape.

Lafala is not the only fictional seafarer to lose his legs. The most infamous fictional sea captain of all, Melville's Ahab, is another amputee. According to Ishmael, *Moby-Dick*'s eloquent narrator:

> [Ahab's] overbearing grimness was owing to the barbaric white leg upon which he partly stood. It had previously come to me that this ivory leg had at sea been fashioned from the polished bone of the sperm whale's jaw. 'Aye, he was dismasted off Japan,' said the old Gay-Head Indian once; 'but like his dismasted craft, he shipped another mast without coming home for it. He has a quiver of 'em.'[93]

Melville's Ahab had been dismasted – rendered impotent – by losing his leg. The prosthetic leg on which he stood was forged from the jawbone of his nemesis species, a sperm whale.

Worker's glove in Khor Fakkan

There are many pointed references to whiteness in *Moby-Dick*, this parable of US empire, most prominently the *white* whale Ahab seeks. In the aforementioned passage, the story of the *white* leg is embellished by an indigenous American who laments Ahab's profligacy with his multiple whalebone false legs. Does losing his sea leg and acquiring a "barbaric white leg" transform Ahab into the megalomaniacal force that proves fatal, not only to the inhabitants of the deeps but also to his worldly crew of seamen? What would have happened if Ahab had chosen to lend an ear and a hand to his sailors, rather than lead them into the deep?

A HAND

In September 2022, two-and-half years after my book *Sinews* was published, I received an email from a Dutch reporter, Arjen van Veelen, who wrote that some of the stories in the book had appealed to him, especially one describing a hunger ship (recounted above on page 48). He wrote:

> Most striking for me – as a Rotterdam-born-and-raised writer – was learning about the revolt on the MV *Saudi Independence* in 1981. It is indeed symptomatic, as you wrote, that the Filipino strikers were erased from history. After reading this story, I contacted a former union worker from Rotterdam, Nico Sannes, who spent a couple of weeks on board of the *Saudi Independence* during the strike. However, even he could not provide me with any of their names let alone [their] later whereabouts. All he could tell me was that the Filipino crew had to leave the ship immediately after the Dutch court dismissed the appeal. They were originally supposed to fly back to the Philippines on a KLM-flight via Jedda. To avoid possible persecution he changed their flights into a direct flight to the Philippines. He did send me a copy of a privately printed memoir with, on its cover, a photo with part of the crew taken on board of the ship (see below). I thought it might interest you in case you hadn't already seen it (photographer unknown to me).

In this extraordinary photograph, Sannes stands amid sixteen of the striking seafarers and the woman friend of one. They are bedecked in early 1980s fashion: one wears a Bjorn Borg-style headband, others sport long hair or mullets. Two dogs sit obediently with the crew.

Van Veelen also sent a link to a piece he had written about Sannes, in which the elderly firebrand recounted how he lost his father in the 1940 German invasion of the Netherlands.[94] His father, a victim of the blistering bombing of Rotterdam, was listed as missing, and his family received no support then or after the war. As he told Van Veelen, he "literally lived off the rubbish dump."[95] Sannes became a seafarer at sixteen on a coal barge on the Irish sea. His time at sea was followed by activism in the union.[96] His internet presence, wholly in Dutch, was predominantly about his union militancy. Finally, I hit upon his story of the MV *Saudi Independence*, published in Dutch on a website dedicated to the life and geography of labour in the Netherlands.[97]

Sannes's memoir provides more detail. At the time an officer of the FNV Transport Union, he boarded the ship to act as a sort of shield, preventing tugboats called in by the Greek captain from pushing the ship away from the port. He remained onboard for weeks, and developed deep friendships with the striking seafarers. He called the tugboat company and ensured that they would not engage with the ship. The shipowner turns out to have been a member of the Saud family, and he eventually hired two carloads of Belgian thugs to forcibly remove Sannes and his comrades from the ship. Sannes drove an old cart he found on the docks into their Mercedes, and when the men tried to intimidate him, he called their bluff by reminding them that "200 dockers waiting for a ship in the canteen" would back him up. While the strike went on aboard *Saudi Independence*, the owner of the ship ordered Greek officers to cut back all sustenance for the seafarers. But this move was matched by dockers joining up with a Filipino seafarer support group to keep feeding the twenty-five men for the next few weeks. As the strike went on, tugboat captains, barge skippers, and assorted dockworkers stepped up to support the striking men. Orri's initial case and the appeal were denied by lower courts, but eventually the Dutch Court

Nico Sannes and the Filipino crew of MV *Saudi Independence*; courtesy of Nico Sannes

of Cassation ruled for the shipowner. Sannes recounts: "Even before the verdict had been fully absorbed, a replacement team came onboard. They were a bunch of unsavoury booze-smelling expired English sailors." Sannes's last act of solidarity was to rebook the men on a flight that would wholly avoid Saudi airspace so that they could evade certain physical punishment – lashings – promised them by the shipowner for their impudence.

Though he had the men's names on the roster, Sannes didn't think these were their real names, as often seafarers – especially those engaged in acts of resistance – carried multiple identities to evade blacklisting. He wept as he saw them to the airport. It was unsurprising that the court of law would enforce the contract and rule on behalf of a Saudi enterprise owned by a member of the royal family, especially as the Dutch government deepened its trade relations with Saudi Arabia. Van Veelen rounds out the narrative:

> His own union was eventually no longer happy with Sannes's solidarity actions. 'Are you going on another outing?' colleagues mocked him. In the long run, Sannes was even forbidden to support strike ships: he would endanger the union with lawsuits.
>
> That still stings him. Solidarity is often reciprocal these days, says Sannes. Like: 'I support you, so that you support me. But what matters is helping people from your heart. Even if it doesn't immediately help you yourself.' [...]
>
> 'Yes, abuses are being identified,' he says. 'Investigative reports are issued. Experts come and say wonderful things. And then it stops. So I say: put that government under such pressure that this inaction is no longer possible. If you discover a wrongdoing, you have to go against it with everything you've got. Sometimes even outside the law.'
>
> Not with symbolic politics, but with social struggle. He hits his fist on the table.
>
> And Sannes demonstrates, beating the table with his old hands, young again in that moment – as if to impress upon us that we are all in the same boat, just as he literally shared a boat with those sailors from afar back then.[98]

I know more about Sannes than I do about the courageous Filipino strikers of MV *Saudi Independence*. I know his name and a little bit about his past. All I know about the seafarers is that one had a woman friend in port and that, despite their hunger, they cared for a dog on the ship. The remaining photo does show smiling faces, jaunty poses, the way they hold each other, flank each other, put hands upon one another's shoulders, the warmth of their gazes at one another.

THE HEART

On a balmy August day in 2020, the crew of Maersk MT *Etienne* were radioed by the Maltese coast guard and told about a rickety fisherman's boat in distress in the Gulf of Gabès near Tunisia's east coast. After the captain and crew of *Etienne* located the migrants' vessel, they set to work and at first tried to secure the boat alongside their ship. As waves and winds picked up, however, their wooden boat began to give way. Amidst the heaving seas and lashing winds, the ship's crew threw down rope ladders and welcomed twenty-seven migrants – who included, newspaper reports emphasise, "a pregnant woman and a child" – aboard.[99]

In a report later published by Maersk Tankers Company, the ship's Russian captain was quoted as saying, "The moment of embarkation was perilous. As we lowered the rope ladder, we knew there were risks of scuffles, falls, injuries, and even fatalities. Factoring the high freeboard and dire weather conditions, I saw that our chances of successful embarkation were slim. But on a ship, you do not turn your back. Even if you have doubts or are afraid, you carry out your duties."[100] After the migrants boarded safely, they watched in horror as their boat disintegrated in the sea and was washed away.

The ship changed course to its original destination, Malta, and awaited permission from Marsaxlokk to moor and discharge their passengers. But radio silence reigned. The ship also contacted port authorities in Tunisia, who ignored them. Maersk's home state, Denmark, where MT *Etienne* is also registered, was no more forthcoming. Where Malta commented at all, it used public hygiene precautions around COVID-19 as an excuse for the inhumanity of refusing safe harbour to the migrants. The standoff between the ship and states continued for thirty-eight days as food onboard dwindled.

Migrant boat moored to MT *Etienne*; courtesy of Maersk Tankers

In photographs on the Maersk website, *Etienne*'s passengers rest on makeshift mats made of ropes and hoses, ensconced in the ribs of the ship, as Jonah in the belly of the whale.[101] Over the next five-and-a-half weeks, the ship's passengers remained there, seeking shade on deck. The ship was fitted and victualled for twenty-four crew members; now it had more than double that number. As it sat at anchor for the next five-and-a-half weeks, food, fuel, and water slowly diminished. Journalists took boats to photograph migrants and crew on deck. One Sudanese passenger wrote a message and threw it down in a plastic bottle. The note included a phone number in Sudan and read: "I'm Gasim… just to confirm my family and my friend that I'm still a life."[102] A life, passed over by billows and waves and rejected at shore.

Migrants aboard MT *Etienne*; courtesy of Maersk Tankers

As the impasse continued, the sense of desperation and captivity intensified. A month into the rescue, three of the new passengers jumped overboard. One crew member who saw them jumping launched a rescue boat. He later said, "Once we had successfully recovered them from the sea, we realized just how desperate we were for outside help." No help ever arrived.[103]

The men who jumped left behind notes for the captain and the crew written on paper plates and coffee filters. One note said:

Dear Captain, We thank you for all your best helpful and we really appreciate you a lot. You show kind of solidarity. You are our hero. Now we need you to send our message to European and humanitarian NGO that in these days if there is no solution about us we will go back to water by any way possible, because Maersk tanks find us inside the waters and became a crime because you safe our life. European don't need us alive.[104]

The note on the coffee filter was from Gasim, the Sudanese migrant who penned the message in the bottle thrown to reporters. This one was addressed to "all the people in the ship" and thanked the crew "for giving us a second chance to achieve our goals."

The horror of the migrants' desperation and the corrosive indifference of all responsible states finally drove the captain to call on *Mare Jonio*, a rescue vessel operated by Mediterranea Saving Humans NGO. He asked it to dock alongside, and for Mediterranea's medics to assess the physical and psychic condition of the migrants. The medics found that the travellers all suffered deteriorating health conditions; *Mare Jonio* boarded the twenty-seven migrants and transferred them to Sicily. When they finally arrived on land, it had been forty days since their rescue, more than ten times longer than Jonah's biblical ordeal at sea.

The operators of *Mare Jonio* were prosecuted by Italian authorities in 2021 for bringing the migrants ashore.

Migrant notes aboard MT *Etienne*; courtesy of Maersk Tankers

DRAMAS & DERMIS

Samia Khatun's meditation on South Asian migration to Australia, *Australianama*, traces the "Seven Voyages of Khawajah Muhammad Bux." Bux was a Punjabi seafarer circulating through the Indian Ocean and the further reaches of the British Empire in the late nineteenth century. Khatun tells us how Bux's travel dramaturgy – self- consciously or not – echoed the seven voyages of Sindbad, the sailor of the *Thousand and One Nights*. The stories Bux brought back from travels to Basra, Mecca, Berbera, and onwards to Shanghai, London, Melbourne, and Fremantle were for the edification and education of those around him, and ultimately served as lessons in human improvement for his heirs. On his first voyage, Bux was brought onboard by a *serang* (recruiter and bosun), and served as a fireman shovelling coal into the steamship's engine. In a massive storm on the way from Bombay to Basra, the ship heaved so badly that the captain forced the seasick firemen to tie ropes around their waists and to continue to feed the maw of the ship. When three new seamen refused, they were beaten so badly that all three attempted suicide by jumping into the sea. One did not survive.

Subsequent journeys saw Bux negotiating a pilgrimage to Mecca through trading goods *en route*; engaging in currency arbitrage; eventually becoming a *serang* himself (overseeing more than 150 deckhands); and being denied entry into Australia, which strictly enforced the colour line not only against migrants, but even seafarers touching its post. Each of Bux's journeys culminated in his return to Bombay's Bhindi Bazar with his tales of travels, travails, and triumphs.[105]

Khatun shows how these sailors' yarns were not only about the journey, or even about backbreaking, soul-destroying work, but also about what she calls "remittance economies," as well as the incongruence between "the physical circulation of sailors" and "the economic circulation of sailors' capital" in the Indian Ocean.[106] They were moral tales – stories about improvements to one's personhood and fortunes. Bux's silences (notably about Somali and Aboriginal peoples he met) and derision (often for South Asian women's practices of homo-sociality) are also telling; they map (dis)identifications and (in)visibilities of entire categories of people in these stories. Bodies, gendered and racialised, beaten and starved, are at the heart of these yarns. As are communities: at sea, at the bazar, in sailortowns.

Before the tightening of borders and demands for ever more paperwork at Atlantic ports from the early 1900s, seafarers jumped ship and went on to take jobs in factories, congregated in neighbourhoods where sailors' boarding houses proliferated, and transformed areas near ports into lively, cosmopolitan enclaves. Seafarers took with them international connections, new and unlikely languages, and news from distant ports. Alan Villiers, who sailed on a *dhow* from the Arabian Peninsula to East Africa, describes a "curious *Yemeni* sayyid who was thinking of going to the Congo to collect some dues," and who seemed to speak a language Villiers could not understand. It turned out that he had been a sailor working the stokehold of steamships. He jumped ship in New York, and made his way to Hamtramck, an urban enclave in Detroit, Michigan, where he worked alongside Polish residents in an automobile factory. The language he was speaking was Polish. He had then returned to settle in Yemen, but his networks and connections reached across the seas.[107]

As far back as the eighteenth century, Wapping, Shadwell, and Whitechapel in London had been destinations for sailors jumping ship. Despite the horrified reaction of Christian missionaries ministering to wayward seafarers in these neighbourhoods, they and the wildly cosmopolitan residents therein thrived.[108] Some married local women picked up their lovers' languages with ease. The first Arab communities of Britain comprised Yemeni seafarers of South Shields and Cardiff.[109] Harlem and Hamburg both attracted communities of lascars who had jumped ship.[110] Claude McKay's novels cited above similarly drew on his experience of tramp ships to describe the bawdy and boisterous Old Port area of Marseille, portraying a cast of characters of African, Caribbean, Arab, and African American seafarers whose itinerant and transgressive lives challenged forms of discipline and containment enforced against them.[111]

As turnaround times at ports diminish, so do spaces for this kind of encounter. But seafarers still drink together or have karaoke sessions; increasingly, with the spread of satellite internet onboard, they remain connected to the world by internet. They tell stories on Tik Tok and Facebook in a panoply of maritime languages. Their stories are contoured to the shape of their homes, lives, loves, and the many paradoxes of life at sea.

If the sailors' yarn is one way that stories of seafaring travel, seafarers' tattoos are another. There is some dispute as to how tattoos came to be indelibly associated with seafaring.[112] Tattoos may have been marks of punishment borne by convicts, who were often pressed into service aboard ships at ports in Europe and in the colonies. The British in India initially punished Brahmin convicts by tattooing them on the forehead and transporting them "over the water" which was considered an abomination; they extended the practice of inscribing crimes on the forehead skin to all Indian convicts in 1795.[113] Or, sailors' tattoos may have been borrowed from indigenous peoples of the South Seas or Oceania, encountered in colonial trade expeditions.[114] Anchors, roses, names of the beloved, and other marks and

Departing Khor Fakkan

tattoos distinguish the seafarer in public. They are pressed into flesh with pigments, historically made of char, soot or gunpowder, today with tattooing ink in more or less hygienic parlours. Seafarers' tattoos code their itineraries, special milestones, and professional achievements, and act as symbols of faith or talismans against malevolent weather and baleful tides. The most beloved character in *Moby-Dick*, the harpoonist Queequeg, was inked with "a complete theory of the heavens and the earth" and "a mystical treatise on the art of attaining truth". As he prepared for his death, he copied these marks onto a coffin he constructed aboard the doomed *Pequod*.[115]

Words and flesh.

Bodies are tied to stories, but also to places. Stories allow people to take their places, wherever these may be, across the seas. But seafarers also see their mobility conditioned, constrained, and defined by racialisation.[116] The ways they are handled (but also what they handle), how they bear an injury, how they walk, how they congregate and fight, the way they are seen (but also how they see), are contoured by their identification papers and the language they speak. That is, by how their skin – a palimpsest of pigment, sunleathering, work-wear, and ink – is seen.

Racialisation, as the scholar Utz McKnight has written in a discussion of Melville, is a way to forge, to make, and to fabricate, "the limitations of our sensory and conceptual ability."[117] The self-fashioning of the flesh and the eyes is one way to escape the enforced enclosures of the sensorium. But all instances of self-fashioning are also constrained by the political economy of exploitation and the global colour line. What we become is not solely what we desire, but often a negotiation, a struggle, between our wishes and the obstructions of class and racialisation. Joy, skill, and experience enter into the making of seafarers' magical bodies. But the catastrophe of capitalism is the ease with which it co-opts situated knowledges and emotional attachments, stripping them of particularity and context, making them into something generic, contained, and quantifiable. Think, for example, of the transformation of karaoke sociality into more private family contact; think of the establishment of indices of happiness – that ultimate capitalist guarantor of productivity – to forestall madness at sea.

Flesh and words, bios and logos. The great Caribbean thinker Sylvia Wynter celebrated the body, its magic, its "muscle and bone animated by hope and desire."[118]

We are at once confined within our finite flesh, straining for the Polestar, *and* we are unbound: across a place, across the sea, across our bodies' skin and sinews. Stories require communities, listeners, audiences. They require a stage, a bazar, a café, a ship's deck where you sit and ravel and unravel rope. They are a thing of collectives.

Judith Butler's classic *Bodies That Matter* reminds us that, "Not only [do] bodies tend to indicate a world beyond themselves, but this movement beyond their own boundaries, a movement of boundary itself, [appears] to be quite central to what bodies 'are.'"[119] Seafarers cross these boundaries physically and metaphorically, again and again. The seafarer's body, the corporeal life of seafaring, is a distillation of the multitudinous paradoxes of capital. The multiple futures and forking paths I have sketched here demand a recognition of the suffering and exploitation of seafarers *and* their capability, their determination, and the durability, suppleness and richness of their lives.

NOTES

1   'United Nations Conference on Trade and Development', *Review of Maritime Transport 2021* (Geneva: UNCTAD, 2021), p. xx.

2   Leon Trotsky, as quoted in Allan Sekula, *Fish Story* (London: MACK, 2018), p. 121; Janet Ewald, 'Roundtable: Reviews of Gopalan Balachandran, *Globalizing Labour? Indian Seafarers and World Shipping, c. 1870-1945* with a Response by Balachandran', *Internation al Journal of Maritime History* 25:1 *(2013)*, 278.

3   Malcolm Lowry, *Ultramarine: A Novel* (London: Penguin Modern Classics, 2000 [1962]), p. 19.

4   'Voyage of the Clipper-ship Great Republic', *New York Times* (6 July 1858), p. 3.

5   Sekula, *Fish Story,* p. 50.

6   Howard Eiland and Michael Jennings, *Walter Benjamin: A Critical Life* (Cambridge, MA: Harvard University Press, 2014), p. 241.

7   Vilhelm Aubert, *The Hidden Society* (Totowa, NJ: Bedminste Press, 1965). Anthropologist Johanna Markkula has rightly criticised me about the categorical nature of stating that ships are total institutions. Her own work, some of which is cited below, shows the incompleteness of captivity aboard ships. But there are tensions and difficulties that remain in work at sea without sight of shore for months on end; I have left this reference to Aubert's argument, but as a question rather than as an affirmation.

8   Aubert, *The Hidden Society.*

9   Vivek Bald, *Bengali Harlem and the Lost Histories of South Asian America* (Cambridge, MA: Harvard University Press, 2013).

10   Alan Villiers, *Sons of Sinbad: Sailing with the Arabs in their Dhows* (London: Hoddern & Stoughton Ltd., 1966); Johan Mathew, *Margins of the Market: Trafficking and Capitalism across the Arabian Sea* (Berkeley, CA: University of California Press, 2016); Yacoub Yusef al-Hijji, *Kuwait and the Sea: A Brief Social and Economic History* (Cowes, Isle of Wight: Arabian Publishing Ltd, 2010); William Lancaster and Fidelity Lancaster, *Honour is in Contentment: Life Before Oil in Ras Al-Khaimah (UAE)* and *Some Neighbouring Regions* (Berlin: de Gruyter, 2011).

11   George Hourani, *Arab Seafaring in the Indian Ocean in Ancient and Early Medieval Times* (Princeton, NJ: University Press, 1995), p. 113.

12 Marcus Rediker, *Between the Devil and the Deep Blue Sea: Merchant Seamen, Pirates, and the Anglo-American Maritime World, 1700–1750* (Cambridge: University Press, 1987); see also Rediker, *Outlaws of the Atlantic: Sailors, Pirates, and Motley Crews in the Age of Sail* (Boston: Beacon Press, 2014).

13 Ray Costello, *Black Salt: Seafarers of African Descent on British Ships* (Liverpool: University Press, 2014); Gopalan Balachandran, *Globalizing Labour? Indian Seafarers and World Shipping, c. 1870–1945* (Oxford: University Press, 2012); Janet Ewald, 'Crossers of the Sea: Slaves, Freedmen and Other Migrants in the North-western Indian Ocean, c. 1750–1914', *American Historical Review* 105:1 (February 2000), 69–91. This essay does not touch on the horrors of the transportation of enslaved humans aboard ships. For this subject, see, *inter alia*, Sowande' Mustakeem, *Slavery at Sea: Terror, Sex, and Sickness in the Middle Passage* (Champaign, IL: University of Illinois Press, 2016); Julius Scott III, *The Common Wind: Afro-American Currents in the Age of the Haitian Revolution* (London: Verso, 2018); Marcus Rediker, *The Slave Ship: A Human History* (London: John Murray, 2018); Marcus Rediker and Peter Linebaugh, *The Many-Headed Hydra: The Hidden History of the Revolutionary Atlantic* (London: Verso, 2002).

14 Bald, *Bengali Harlem*; Balachandran, *Globalizing Labour*; Ewald, 'Crossers of the Sea'.

15 Dick Lawless, 'The Role of Seamen's Agents in the Migration for Employment of Arab Seafarers in the Early Twentieth Century', *Immigrants and Minorities* 13:2–3 (1994), 35–58.

16 Rodney Carlisle, *Sovereignty for Sale: The Origins and Evolution of the Panamanian and Liberian Flags of Convenience* (Annapolis, MD: Naval Institute Press, 1981).

17 Rodney Carlisle, 'The "American Century" Implemented: Stettinius and the Liberian Flag of Convenience', *Business History Review* 54:2 (1980), 175–91.

18 Quoted in L. F. E. Goldie, 'Environmental Catastrophes and Flags of Convenience - Does the Present Law Pose Special Liability Issues?' *Pace International Law Review* 3:1 (1991), 11 n. 47.

19 'United Nations Conference on Trade and Development', *Review of Maritime Transport 2022* (Geneva: UNCTAD, 2022), pp. 41–42.

20 'United Nations Conference on Trade and Development', *Review of Maritime Transport 2021* (Geneva: UNCTAD, 2021), p.xx.

21 V. C. Burton, 'Counting Seafarers: The Published Records of the Registry of Merchant Seamen 1849–1913', *Mariner's Mirror* 71:3 (1985), 317.

22 C. L. R. James, *Mariners, Renegades and Castaways: The Story of Herman Melville and the World We Live In* (Dartmouth: College Press, 2001), p. 26.

23 James, *Mariners*, p. 13.

24 Deborah Cowen, *The Deadly Life of Logistics: Mapping Violence in Global Trade* (Minneapolis: University of Minnesota Press, 2014). See also Rafeef Ziadah, 'Circulating Power: Humanitarian Logistics, Militarism, and the United Arab Emirates', *Antipode* 51:5 (2019), 1684–1702, for the spillage of these logics into the adjacent fields of humanitarianism.

25 Vanessa Agard-Jones, 'Bodies in the System', *Small Axe* 42 (2013), 182–92.

26 Rafael Lefkowitz and Martin Slade, *Seafarer Mental Health Study* (London: ITF Seafarers' Trust and Yale University, 2019), pp. 20–21.

27 Leon Fink, *Sweatshops at Sea: Merchant Seamen in the World's First Globalized Industry, from 1812 to Present* (Chapel Hill: University of North Carolina Press, 2011).

28 I owe this insight to discussions with Alishba Zaman.

29 Swire International is a shipping conglomerate whose origins can be traced to John Swire and Sons Limited, established in Liverpool in 1816 and engaged in colonial enterprise in the Pacific.

30 Monthly Station Report from Dubai, date presumed to be 1985. The Archive of Mission to Seafarers; Hull History Centre. The document is undated, and as the ship is not named, the exact date had to be triangulated through other discussions in the same report as well as through reports from Bahrain.

31 Frank Broeze, 'The Muscles of Empire — Indian Seamen and the Raj 1919-1939', *Indian Economic and Social History Review* 18:1 (1981), 46.

32 Rediker, *Between the Devil and the Deep Blue Sea*, p. 50.

33 McKinsey & Company, Inc. for the British Transport Docks Board, *Containerization: The Key to Low-Cost Transport.* (London: British Transport Docks Board, 1967), pp. 3–4.

34 McKinsey & Company, *Containerization*, p. 5.

35 McKinsey & Company, *Containerization*, p. 16.

36 Rediker, *Between the Devil and the Deep Blue Sea*, p. 95.

37 Burton, 'Counting Seafarers', p. 316.

38 Balachandran, *Globalizing Labour*, p. 108–09. See also Bald, *Bengali Harlem*, p. 104.

39 Rediker, *Outlaws of the Atlantic*, pp. 13–14

40 Herman Melville, *Moby-Dick, or, the Whale* (London: Penguin Books, based on Northwestern University Press edition, 1992), p. 236.

41 Villiers, *Sons of Sinbad*, p. 30.

42   See also Johanna Markkula, '"Any Port in a Storm": Responding to Crisis in the World of Shipping', *Social Anthropology* 19:3 (September 2011), 297–304.

43   Kale Fajardo, *Filipino Crosscurrents: Oceanographies of Seafaring, Masculinities, and Globalization* (Minneapolis: University of Minnesota Press, 2011), p. 106.

44   Vilhelm Aubert and Oddvar Arner, 'On the Social Structure of the Ship', *Acta Sociologica* 3:4 (1959), 204.

45   B. Traven, *The Death Ship: The Story of an American Sailor* (Chicago: Review Press, 1991 [1934]), p. 150.

46   Balachandran, *Globalizing Labour*, p. 123.

47   Broeze, 'Muscles of Empire', p. 46.

48   William Hunter, *An Essay on the Diseases Incident to Indian Seamen, or Lascars on Long Voyages* (Calcutta: The Honorable Company's Press, 1804), p. 12.

49   Aaron Jaffer, '"Lord of the Forecastle": Serangs, Tindals, and Lascar Mutiny, c. 1780–1860', *International Review of Social History* 58 (2013), 153–75.

50   Costello, *Black Salt*, p. 128.

51   Letter from Captain C. K. Gabriel to A. G. Selander, Deputy General Secretary of ITF (26 August 1981); ITF Seafarers Section; Various Correspondence Relating to the Arab Countries and the Arab Shipping Company, September 1978–September 1982. (MSS.648/SF/2/2/45). Warwick University Trade Union Archives.

52   Richard David Poisson, 'Seafarers and International Shipping Standards', Unpublished Masters Dissertation (University of Rhode Island, 1982), Appendix, p. 10.

53   M. Muller, 'Strike of Crew Members Supported by ITF Held Unlawful by Dutch Court Applying Philippine Law', *Journal of Maritime Law and Commerce* 16:3 (1985), 423–26

54   Laleh Khalili, *Sinews of War and Trade: Shipping and Capitalism in the Arabian Peninsula* (London: Verso, 2020), pp. 240–41.

55   'Mission to Seafarers Monthly Port Report (5 July 1993)', Box 280, Missions to Seafarers Dubai (uncatalogued as of 7 November 2019); Hull History Centre Archives.

56   Karen McVeigh, 'Abandoned at Sea: The Crews Cast Adrift Without Food, Fuel or Pay', *The Guardian* (12 April 2019).

57   Karen McVeigh, 'Seafarers Trapped on Ship for 33 Months Say Jail Threats Forced Them to Reboard', *The Guardian* (11 July 2019).

58   Nick Webster, 'Abandoned UAE Sailors Still Owed More Than Dh500,000 Weeks After Rescue', *The National* (1 September 2019).

59   'Mission to Seafarers Monthly Port Report (3 December 1993)'
     Box 280, Missions to Seafarers Dubai (uncatalogued as of 7
     November 2019); Hull History Centre Archives.

60   'UAE: Case of Abandoned Ship, Azraqmoiah, Comes to Close
     After Crew Repatriated After Two Years', *Business and Human
     Rights* (21 June 2019) <https://www.business-humanrights.org/en/
     uae-case-of-abandoned-ship-azraqmoiah-comes-to-close-after-
     crew-repatriated-after-two-years-0> [accessed 25 June 2023].

61   Amitav Ghosh, as quoted in Ravi Ahuja, 'Capital at Sea, Shaitan
     Below Decks? A Note on Global Narratives, Narrow Spaces, and
     the Limits of Experience', *History of the Present* 2:1 (2012), 78.

62   Ahuja, 'Capital at Sea', 82.

63   Gholam-Hossein Sa'edi, *Ahl-e-Hawa (People of the Spirit)* (Tehran:
     Amir Kabir Publishing, 1976), p. 47, my translation.

64   Sa'edi, *Ahl-e-Hawa*, p. 47.

65   George Horne, 'The Tanker -- Queen of the Seas', *New York Times*
     (23 September 1956).

66   Ibid.

67   Nikos Kavvadias, 'Letter from Marseille' in *The Collected Poems of
     Nikos Kavadias,* trans. Gail Holst-Warhaft (River Vale, NJ: Cosmos
     Publishing, 2006), p. 79.

68   Kavvadias, 'Fatas Morgana', *The Collected Poems*, p. 203.

69   Lefkowitz and Slade, *Seafarer Mental Health Study*, p. 5.

70   Fajardo, *Filipino Crosscurrents,* p. 106.

71   UK P&I Club, *Risk Focus: Mental Health* (1 February 2016)
     <https://www.ukpandi.com/news-and-resources/publications/risk-
     focus-mental-health> [accessed on 25 June 2023].

72   Stephen Roberts, et al., 'Suicides Among Seafarers in UK
     Merchant Shipping, 1919–2005', *Occupational Medicine* 60 (2010), 57.

73   Lefkowitz and Slade, *Seafarer Mental Health Study*, p. 5.

74   Declan Bush, 'Suicides at Sea Go Uncounted as Crew Change
     Crisis Drags On', *Lloyd's List* (22 February 2021).

75   Ana Sliškovič, 'Seafarers' Well-being in the Context of the COVID-19
     Pandemic: A Qualitative Study', *Work* 67 (2020), 804.

76   Broeze, 'Muscles of Empire'; Marika Sherwood, 'Race, Nationality
     and Employment among Lascar Seamen, 1660 to 1945', *Journal of
     Ethnic and Migration Studies* 17:2 (1991), 229–244; Marika
     Sherwood, 'Lascar Struggles against Discrimination in Britain
     1923–45: The Work of N. J. Upadhyaya and Surat Alley', *Mariner's
     Mirror* 90:4 (2004), 438–55; Heather Goodall, 'Uneasy Comrades:
     Tuk Subianto, Eliot V. Elliott and the Cold War', *Indonesia and the
     Malay World* 117 (2012).

77 Marcel van der Linden, 'Workers Who Benefit from the Exploitation of Other Workers', *Revista Latinoamericana de Trabajo y Trabajadores 1* (2020). Emphasis in the original.

78 Laleh Khalili, *Heroes and Martyrs of Palestine: The Politics of National Commemoration* (Cambridge: University Press, 2007), p. 224.

79 James, *Mariners*, p. 35.

80 Jorge Luis Borges, 'The Garden of Forking Paths', in *Collected Fictions*, trans. Andrew Hurley (New York: Viking, 1998), p. 125.

81 In a short lecture given in 1967, Michel Foucault describes Persian gardens, cemeteries, boarding schools, colonial missions, brothels, and oceangoing ships as heterotopian spaces "which are linked with all the other [spaces and sites around us], which however contradict all the other sites." His notion of the ship as heterotopian is at once suggestive, imaginative, and problematic. He describes ships as equivalent to colonies or brothels, as instruments of economic development, but also as sites of dreaming and aspiration. The essay elides the violence and the power hierarchies aboard ships and in colonies (and in brothels too). But there is also an odd truth to the idea of the ship as an instrument of economic development, insofar as "economic development" has long been a euphemism for brute force, dispossession, and exploitation. Michel Foucault, 'Of Other Spaces: Utopias and Heterotopias', trans. Jay Miskowiec, *Architecture / Mouvement / Continuité* (October 1984 (1967)), 1–9.

82 Andrea Peterson, 'Why Naval Academy Students are Learning to Sail by the Stars for the First Time in a Decade', *Washington Post* (17 February 2016).

83 Hasan Salih Shihab, 'Stellar Navigation of the Arabs', in Anthony Constable and William Facey (eds), *The Principles of Arab Navigation* (Arabian Publishing Ltd, 2013), p. 25.

84 Margaret Schotte, 'Sailors, States, and the Creation of Nautical Knowledge', in Lauren Benton and Nathan Perl-Rosenthal (eds), *A World at Sea: Maritime Practices and Global History* (Philadelphia: University of Pennsylvania Press, 2020).

85 Raymond Williams, *The Country and the City in the Modern Novel* (Oxford: University Press, 1975), p. 289.

86 Rai Ocampo, posting as the administrator of the Lakbay Kabaro group on Facebook (2 July 2023), <https://www.facebook.com/watch/?v=1500053987470037> [accessed 7 July 2023].

87 Honoré Balzac, *Theory of Walking*, quoted in S. Collado-Vázquez and J. M. Carrillo, 'Balzac and Human Gait Analysis', *Neurologia* 30:4 (2015), 245.

88 Thomas Stoffregen, et al., 'Getting Your Sea Legs', *PLOS ONE* 8:6 (2013), 1–16.

89   Timothy Hain, et al., 'Mal de Débarquement', *Archives of Otorhinolaryngology: Head & Neck Surgery* 125 (1999), 613–20.

90   Claude McKay, *Banjo: A Story Without a Plot* (New York: Ecco Press, 1970), p. 11; Peter Olsen, 'The Negro Maritime Worker and the Sea', *Negro History Bulletin* 34:2 (1971), 40.

91   Claude McKay, *Romance in Marseille*; edited and with an introduction by Gary Edward Holcomb and William J. Maxwell (London: Penguin Classics, 2020), p. 4.

92   McKay, *Romance in Marseille*, p. 19. All caps in the original.

93   Melville, *Moby-Dick*, p. 124.

94   I tried to contact Sannes by email but did not receive a response, likely because I wrote in English.

95   Arjen van Veelen, 'Op de vuist voor betere arbeidsomstandigheden: wat we kunnen leren van Rotterdamse havenwerkers' ('Fighting for Better Working Conditions: What We Can Learn from Rotterdam Dock Workers'), *de Correspondent* (25 November 2022); <https://decorrespondent.nl/13973/op-de-vuist-voor-betere-arbeidsomstandigheden-wat-we-kunnen-leren-van-rotterdamse-havenwerkers/09da87c7-dd78-0a26-112f-c572fade1a85> [accessed 30 June 2023].

96   Van Veelen, 'Fighting for Better Working Conditions'.

97   Nico Sannes, *De Bitter Zee: "Saudi Independence"* (*The Bitter Sea: Saudi Independence*; Rotterdam: Arbeid en Werkstad, 2023), <https://vrijeuitgeverij.nl/wp-content/uploads/2023/04/DE-BIT-TERE-ZEE-1.pdf> [accessed 30 June 2023].

98   Van Veelen, 'Fighting for Better Working Conditions'.

99   Megan Specia, 'Tanker Asked to Rescue Migrants Off Malta Is Denied Permission to Dock', *New York Times* (4 September 2020).

100   Qandeel Shaam, 'I Enjoyed Being the Commander of Such a Heroic Crew During the Rescue Operation', (7 December 2020), posted at <https://maersktankers.com/newsroom/captain-yeroshkin-on-his-crews-rescue-and-care-of-27-persons-caught-at-sea-for-38-days> [accessed 9 June 2023].

101   The pictures can be found at <https://maersktankers.com/newsroom/captain-yeroshkin-on-his-crews-rescue-and-care-of-27-persons-caught-at-sea-for-38-days> [accessed 2 July 2023].

102   Karl Azzopardi, '"Please Tell My Family I Am Alive" – Maersk *Etienne* Migrant's Message in a Bottle', *Malta Today* (8 September 2023), <https://www.maltatoday.com.mt/news/national/104612/please_tell_my_family_i_am_alive_maersk_etienne_migrant> [accessed 2 July 2023].

103   Qandeel Shaam, 'I Enjoyed Being the Commander'.

104 Twitter post by @MaerskTankers on 4 September 2020 <https://twitter.com/MaerskTankers/status/1301859674967478273> [accessed 8 June 2023].

105 Samia Khatun, 'Seven Voyages of Khawajah Muhammad Bux', in *Australianama: The South Asian Odyssey in Australia* (London: Hurst, 2018).

106 Khatun, *Australianama*, p. 79; p. 67.

107 Villiers, *Sons of Sinbad*, p. 45; pp. 64–65.

108 Joseph Salter, *The Asiatic in England: Sketches of Sixteen Years' Work Among Orientals* (Limehouse, UK: Seeley, Jackson and Halliday, 1873); *The East in the West, or, Work among the Asiatics and Africans in London* (London: S.W. Partridge & Co, 1895).

109 Fred Halliday, *Arabs in Exile: Yemeni Migrants in Urban Britain* (London: I. B. Tauris, 1992); Mohammed Sidiq Seddon, *The Last of the Lascars: Yemeni Muslims in Britain 1836–2012* (Markfield, UK: Kube Publishing, 2014).

110 Bald, *Bangali Harlem*.

111 McKay, *Banjo*; *Romance in Marseille*.

112 Rediker, *Between the Devil and the Deep Blue Sea*, p. 12.

113 Radhika Singha, 'Settle, Mobilize, Verify: Identification Practices in Colonial India', *Studies in History* 16:2 (2000), 151–98.

114 Les Back, 'Inscriptions of Love', in Helen Thomas and Jamilah Ahmed (eds) *Cultural Bodies: Ethnography and Theory* (London: Blackwell Publishing, 2004), 27–54.

115 Melville, *Moby-Dick*, p. 524.

116 Steven C. McKay, 'Racializing the High Seas: Filipino Migrants and Global Shipping', in John Park and Shannon Gleeson (eds), *The Nation and Its Peoples. Citizens, Denizens, Migrants* (London: Routledge, 2014); Johanna Markkula, '"We Move the World": The Mobile Labor of Filipino Seafarers', *Mobilities* 16:2 (2021), 164–77.

117 Utz McKnight, *Race and the Politics of the Exception: Equality, Sovereignty, and American Democracy* (London: Routledge, 2013), p. 99.

118 Sylvia Wynter, 'The Pope Must Have Been Drunk, the King of Castile a Madman: Culture as Actuality, and the Caribbean Rethinking Modernity,' in Alvina Ruprécht and Cecilia Taiana (eds), *The Reordering of Culture: Latin America, The Caribbean and Canada, In the Hood* (Ottawa: Carleton University Press, 1995), pp. 17–42.

119 Judith Butler, *Bodies That Matter: On the Discursive Limits of 'Sex'* (London: Routledge, 1993), p. ix.

ACKNOWLEDGEMENTS

For allowing me to present earlier versions of this essay and for engaging with it, I am grateful to Utz McKnight at the University of Alabama; Jess Bier at Erasmus University Rotterdam; the British Society for Middle Eastern Studies; the Centre for Postcolonial Studies at Goldsmiths, University of London; Casa Árabe, Madrid; the School of Sociology, Politics, and International Studies, University of Bristol; the Department of Politics and International Studies, Cambridge University; and the generous and exacting folks at the Ninth Ethnography and Qualitative Research International Conference, University of Trento, Italy.

For speaking to me, teaching me, and telling me their stories, I am indebted to the seafarers whose ships I took, and to those who spoke to me via email, social media, or in other settings. I cannot claim to represent even a fraction of their experience at sea and on shore, but I hope to honour their labours.

I have not included in this book any photographs from my journeys of identifiable seafarers, as I never asked for, nor received, consent to reproduce their likeness. Other photographs where seafarers' faces can be seen are taken from publicly available news sources with permission. Unattributed photographs were taken by me.

Unless noted otherwise below, all photographs were taken by Laleh Khalili and are copyright of the author.

Page 31: Photograph by Norbert Schiller
Page 77: Courtesy of Nico Sannes
Pages 82-85: Courtesy of Maersk Tankers

Laleh Khalili
The Corporeal Life of Seafaring

First edition published by MACK
© 2024 MACK for this edition
© 2024 Laleh Khalili for her text

Design by Carlotta di Lenardo
Edited by Jess Gough
Copy-edited by Julian Myers-Szupinska
Printed in Lithuania

ISBN 978-1-915743-26-8
mackbooks.co.uk